WATER HAMMER RESEARCH
Advances in Nonlinear Dynamics Modeling

WATER HAMMER RESEARCH

Advances in Nonlinear Dynamics Modeling

Kaveh Hariri Asli, PhD

Apple Academic Press

TORONTO   NEW JERSEY

© 2013 by
Apple Academic Press Inc.
3333 Mistwell Crescent
Oakville, ON L6L 0A2
Canada

Apple Academic Press Inc.
1613 Beaver Dam Road, Suite # 104
Point Pleasant, NJ 08742
USA

First issued in paperback 2021

Exclusive worldwide distribution by CRC Press, a Taylor & Francis Group

ISBN 13: 978-1-77463-262-8 (pbk)
ISBN 13: 978-1-926895-31-4 (hbk)

Library of Congress Control Number: 2012951944

Library and Archives Canada Cataloguing in Publication

Water hammer research: advances in nonlinear dynamics modeling/edited by Kaveh Hariri Asli.

Includes bibliographical references and index.
ISBN 978-1-926895-31-4
1. Water hammer. 2. Computational fluid dynamics. 3. Water-pipes--Hydrodynamics.
I. Asli, Kaveh Hariri

TC174.W28 2013   620.1'064   C2012-906415-7

Apple Academic Press also publishes its books in a variety of electronic formats. Some content that appears in print may not be available in electronic format. For information about Apple Academic Press products, visit our website at **www.appleacademicpress.com**

About the Author

Kaveh Hariri Asli, PhD

Kaveh Hariri Asli, PhD, is a professor at the Department of Mathematics and Mechanics, National Academy of Science of Azerbaijan. He is also a professional mechanical engineer with over 30 years of experience in practicing mechanical engineering design and teaching. He is author of over 50 articles and reports in the fields of fluid mechanics, hydraulics, automation and control systems. Dr. Hariri consulted for a number of major corporations.

Contents

List of Abbreviations

CM	Condition base maintenance
FD	Finite differences
FE	Finite elements
FV	Finite volume
FVM	Finite volume method
GIS	Geography information systems
MOC	Method of characteristics
PLC	Program logic control
RTC	Real-time control
WCM	Wave characteristic method

List of Symbols

V = Water flow or discharge $\left(m^3/s\right), \left(lit/s\right)$

C = The wave velocity $\left(m/s\right)$

E_{α} = Modulus of elasticity of the liquid (water), $E_{\alpha} = 2\cdot10^9\,Pa\,\left(kg/m^2\right)$

E = Modulus of elasticity for pipeline material Steel, $E = 10^{11}\,Pa,\left(kg/m^2\right)$

d = Outer diameter of the pipe (m)

δ = Wall thickness (mm)

V_0 = Liquid with an average speed $\left(m/s\right)$

T = Time (S)

h_0 = Ordinate denotes the free surface of the liquid (m)

u = Fluid velocity $\left(m/s\right)$

λ = Wavelength

$(hu)_x$ = Amplitude a

$\dfrac{\partial h}{\partial t}dx$ = Changing the volume of fluid between planes in a unit time

h_0 = Phase velocity $\left(m/s\right)$

v_{Φ} = Expressed in terms of frequency

f = Angular frequency

ω = Wave number

Φ = A function of frequency and wave vector

$v_{\partial}(k)$ = Phase velocity or the velocity of phase fluctuations $\left(m/s\right)$

$\lambda(k)$ = Wavelength

k = Waves with a uniform length, but a time-varying amplitude

$k_{**}(\omega)$ = Damping vibrations in length

ω = Waves with stationary in time but varying in length amplitudes

P_{si0} = Saturated vapor pressure of the components of the mixture at an initial temperature of the mixture $T_0, (Pa)$

μ_2, μ_1 = Molecular weight of the liquid components of the mixture

B = Universal gas constant

P_i = The vapor pressure inside the bubble (Pa)

T_{ki} = Temperature evaporating the liquid components $(^{\circ}C)$

l_i = Specific heat of vaporization

D = Diffusion coefficient volatility of the components

N_{k_0}, N_{c_0} = Molar concentration of 1-th component in the liquid and steam

c_l and c_{pv} = Respectively the specific heats of liquid and vapor at constant pressure

a_l = Thermal diffusivity

ρ_v = Vapor density $\left(kg\!\!\left/\!\!{m^3} \right. \right)$

$R = r = R(t)$ = Radius of the bubble (mm)

λ_l = Coefficient of thermal conductivity

ΔT = Overheating of the liquid (^oC)

β = Is positive and has a pronounced maximum at $k_0 = 0,02$

P_1 and P_2 = The pressure component vapor in the bubble (Pa)

P_∞ = The pressure of the liquid away from the bubble σ

σ = Surface tension coefficient of the liquid

ν_1 = Kinematic viscosity of the liquid

k_R = The concentration of the first component at the interface

n_i = The number of moles

V = Volume (m^3)

B = Gas constant

T_v = The temperature of steam (^oC)

$\rho_i^{\prime}$ = The density of the mixture components in the vapor bubble $\left(kg\!\!\left/\!\!{m^3} \right. \right)$

μ_i = Molecular weight

P_{Si} = Saturation pressure (Pa)

l_i = Specific heat of vaporization

k = The concentration of dissolved gas in liquid

v_0 = Speed of long waves

h = Liquid level is above the bottom of the channel

ξ = Difference of free surface of the liquid and the liquid level is above the bottom of the channel (a deviation from the level of the liquid free surface)

u = Fluid velocity $\left(m\!\!\left/\!\!{s} \right. \right)$

τ = Time period

a = Distance of the order of the amplitude

k = Wave number

$v_0(k)$ = Phase velocity or the velocity of phase fluctuations

$\lambda(k)$ = Wavelength

$\omega_{**}(k)$ = Damping the oscillations in time

λ = Coefficient of combination

q = Flow rate $\left(m^3\!\!\left/\!\!{s} \right. \right)$

μ = Fluid dynamic viscosity $\left(kg\!\!\left/\!\!{m.s} \right. \right)$

γ = Specific weight $\left(N\!\!\left/\!\!{m^3} \right. \right)$

j = Junction point (m)

y = Surge tank and reservoir elevation difference (m)

k = Volumetric coefficient $\left(GN/m^2 \right)$

T = Period of motion

A = Pipe cross-sectional area (m^2)

dp = Static pressure rise (m)

h_p = Head gain from a pump (m)

h_L = Combined head loss (m)

E_v = Bulk modulus of elasticity $(Pa), \left(kg/m^2 \right)$

α = Kinetic energy correction factor

P = Surge pressure (Pa)

g = Acceleration of gravity $\left(m/s^2 \right)$

K = Wave number

T_P = Pipe thickness (m)

E_P = Pipe module of elasticity $(Pa), \left(kg/m^2 \right)$

E_W = Module of elasticity of water $(Pa), \left(kg/m^2 \right)$

C_1 = Pipe support coefficient

$Y\max = Max.$ Fluctuation

R_0 = Radiuses of a bubble (mm)

D = Diffusion factor

β = Cardinal influence of componental structure of a mixture

$N_{k_0}, \; N_{c_0}$ = Mole concentration of 1-th component in a liquid and steam

γ = Adiabatic curve indicator

c_l, c_{pv} = Specific thermal capacities of a liquid at constant pressure

a_l = Thermal conductivity factor

ρ_v = Steam density $\left(kg/m^3 \right)$

R = Vial radius (mm)

λ_l = Heat conductivity factor

k_0 = Values of concentration, therefore

w_l = Velocity of a liquid on a bubble surface $\left(m/s \right)$

P_1 and P_2 = Pressure steam component in a bubble (Pa)

P_∞ = Pessure of a liquid far from a bubble (Pa)

σ and V_1 = Factor of a superficial tension of kinematics viscosity of a liquid

B = Gas constant

T_v = Temperature of a mixture $(^\circ C)$

ρ_i' = density a component of a mix of steam in a bubble $\left(kg/m^3 \right)$

μ_i = molecular weight

j_i = the stream weight

i = components from an $(i=1,2)$ inter-phase surface in $r = R(t)$

w_i = diffusion speeds of a component on a bubble surface $\left(m/s\right)$

l_i = specific warmth of steam formation

k_R = concentration 1^{-th} components on an interface of phases

T_0, T_{ki} = liquid components boiling temperatures of a binary mixture at initial pressure $p_0, \left(^{o}C\right)$

D = diffusion factor

λ_l − heat conductivity factor

Nu_l = parameter of Nusselt

a_l = thermal conductivity of liquids

c_l = factor of a specific thermal capacity

Pe_l = Number of Pekle.

Sh = parameter of Shervud

Pe_D = diffusion number the Pekle

ρ = density of the binary mix $\left(kg/m^3\right)$

t = Time (s)

λ_0 = Unit of length

V = Velocity $\left(m/s\right)$

S = Length (m)

D = Diameter of each pipe (mm)

R = Pipe radius (mm)

v = Fluid dynamic viscosity $\left(kg/m.s\right)$

h_p = Head gain from a pump (m)

h_L = Combined head loss (m)

C = Velocity of surge wave $\left(m/s\right)$

P/γ = Pressure head (m)

Z = Elevation head (m)

$V^2/2g$ = Velocity head (m)

γ = specific weight $\left(N/m^3\right)$

Z = elevation (m)

H_P = Surge wave head at intersection points of characteristic lines (m)

V_P = Surge wave velocity at pipeline points- intersection points of characteristic lines $\left(m/s\right)$

V_{ri} = Surge wave velocity at right hand side of intersection points of characteristic lines $\left(m/s\right)$

H_{ri} = Surge wave head at right hand side of intersection points of characteristic lines (m)

V_{le} = Surge wave velocity at left hand side of intersection points of characteristic lines $\left(m/s\right)$

H_{le} = Surge wave head at left hand side of intersection points of characteristic lines (m)

P = pressure $(bar), \left(N/m^2\right)$

dV - Incremental change in liquid volume with respect to initial volume

$\left(d\rho/\rho\right)$ -incremental change in liquid density with respect to initial density

Superscripts

C^- = characteristic lines with negative slope

C^+ = characteristic lines with positive slope

Subscripts

Min. =Minimum

Max.= Maximum

Lab.=Laboratory

Preface

Water transmission failure sometimes happens due to unusual factors that can suddenly change in the boundaries of the system. High surge pressure during an earthquake, the pump power goes off and lets in air by the air inlet valve, a high discharge rate due to connections and consumers are some of the unusual factors that can suddenly change the boundaries of the system. Most of the transients in water and wastewater systems are the result of changes in the properties and boundaries of the system. These changes can generate the spread of the surge wave and change liquid's properties in pipes and channels. It causes the formation and collapse of vapor bubbles or cavitations and air leakage.

In this book a computational and practical method was used for prediction of water transmission failure. I proposed method allowed for any arbitrary combination of devices in a water pipeline system. A scale model and a prototype (real) system were used for a water pipeline. This book presents the performances of the computational method for water hammer prediction by numerical analysis and nonlinear dynamic model. In the book various methods were developed to solve transient flow in pipes. This range includes the approximate equations to numerical solutions of the non-linear Navier–Stokes equations. The model was presented by Eulerian method based expressed in a method of characteristics (MOC). It was defined by a finite difference form for heterogeneous model with varying state in the system. The book offers MOC as a computational approach from theory to practice in numerical analysis modeling. Therefore it is presented as a computationally efficient method for transient flow irreversibility prediction in a practical case.

The book includes the research of the author on the development of an optimal mathematical model. In order to predict water transmission failure and the propagation of the surge pressure in the pipeline numerical experiments were condected to assess the adequacy of the proposed model. The problem was presented by means of theoretical and experimental research. The author also used modern computer technology, and mathematical methods for analysis of nonlinear dynamic processes. This collection develops a new method for the calculation of water hammer by computer technology. The process of entering input for the calculation of water hammer was simplified for the user through the use of Geography Information Systems (GIS). The author used a parametric modeling technique and multiple regression analysis for a water pipeline. This method has provided a suitable way for detecting, analyzing, and recording transient flow (down to 5 milliseconds), hence it was recommended for linking to a Programmable Logic Control (PLC) system. Certainly it can be assumed as a method with high speed response ability for damping the water hammer phenomena during irregular condition. The response time is the time it takes for a pressure wave to travel the pipe system's greatest length two times. The author believes that the results of this book show important information that can help to reduce the risk of system damage or failure at the water pipeline.

— **Kaveh Hariri Asli, PhD**

Introduction

Mixtures of liquids, which are formed due to deep penetration (mixing) in the flow of two fluids between parallel plates or turbulent mixing in the pipes, are prevalent in our everyday lives. Deep penetration of the mixture components can be considered as a combination of diffusion and physical mixing. The first process dominates in the deep penetration and the second stage the liquids are mixed. Immiscible mixing is also crucial in modern technology. It allows chemists to control chemical reactions for the production of polymeric materials with unique properties and distribution of additives that reduce the viscous friction in the pipes. However, despite its popularity, both in nature and in the production, the mixing process is still not completely clear. Researchers in different areas can not yet even set a common terminology for it by using different names. The mixing process is extremely complex and is found in a variety of systems. Theory of mixing is included, for example, soluble and partially soluble, chemically active and inert liquid, slow laminar flows, and rapid turbulent flows. Not surprisingly, there is no single theory capable to explain in detail the process of mixing in fluids. Therefore the direct calculation is impossible to cover all important aspects of this phenomenon. Nevertheless, some information about the process of mixing can be obtained both through physical experiments and using computer simulation. Typically, in certain places or local points of water pipelines exists sharp changes in pressure above atmospheric pressure. If there is a leak in the pipeline or through the valves, spaces in the plumbing lines, is filled with gaseous phase (e.g., steam at ambient temperature) or air. The complex microscopic interaction between the components of liquid-gas mixture makes the simulation extremely important. There have been several special studies by using computer graphics in the light of immiscible mixtures. However, there are few works that deal with blending fluid. Changes in the properties of liquids in pipes and channels are due to factors such as decompression (because of the sudden opening of the discharge valve), the spread of the pulse pressure, heating or cooling, or energy production systems, mixing with the particulate matter or other body fluids (which can change the density of the liquid, specific gravity and viscosity), the formation and collapse of vapor bubbles (cavitation) and air leak or disconnection of the system (near the vent and/or pressure wave). Changes to the boundaries of the system are due to factors such as the rapid opening or closing a valve, pipe explosion (due to high pressure) or the collapse of the tube (due to low pressure), stop the pump inlet air in the vacuum circuit breaker, the penetration of water through the valve, massive outflow valve in the discharge pressure or fire hoses, damage to the disk and / or resonance in the switching valve. Such sudden changes make the transition pressure pulse, which quickly spreads far from the place of origin of perturbations, in any possible direction, and across the sealed system. Most of the transients in water and drainage systems are the result of changes in the boundaries of the system. It happens usually near the end of the system upstream or downstream or in the local high points. Consequently, the results of this book will help reduce the risk of system damage or failure with the proper analysis to provide a dynamic response to the shortcomings of the system, design protection equipment to manage the transition energy and determine the operational procedures to avoid transients. Analysis, design and operating procedures, that is all the benefits of computer simulations, are discussed in the book. Study of hydraulic transients began with the work of Zhukovsky [108] and Allievi [18]. Many researchers have made significant contributions in this area, including Wood F. [163], R. Angus and John Parmakian [129, 130], who popularized and perfected the graphical method of calculation. Benjamin Wylie and Victor Streeter [165-167, 147-149], method of characteristics combined with computer modeling. Subject of transients in liquids are still growing fast around the world.   Brunone et al [24], Koelle and Luvizotto [112], Filion and Karney [36], Hamam and McCorquodale [44], Savic and Walters [140, 141], Walski and Lutes [159], Wu and Simpson [164], have been developed various methods of investigation of transient pipe flow. These ranges of methods are included by approximate equations to numerical solutions of the nonlinear Navier–Stokes equations. Basic theory of unsteady fluid flow in pressure pipelines were set out in the works of Zhukovsky. He obtained the differential equations of motion of inviscid fluid formed the basis development of the theory about pressure and pressure flow of viscous fluid. With the help of this theory, it became to explanation of the physical phenomenon, known as water hammer. N.E. Zhukovsky introduced the concept of the effective sound speed. He mentioned to reducing the motion of a compressible fluid in an elastic cylindrical pipe to the motion of a compressible fluid in a rigid pipe, but with a lower modulus of elasticity of the liquid. Further study of transients in the pipelines pursuits in the works of I.A. Charny, Khristianovich, A.H. Mirzajanzadeh, M.A. Hussein-Zadeh, V.A. Yufina, H.N. Nizamova, R. F. Ganieva, L.B. Kublanovskaya, L. Polyansky, A.K. Galliamova, M.V. Lur'e, E.V. Vyazunova, A.G. Gumerova, A. Shumaylova, A. Kozak, A.A. Kandaurova, E.M. Klimovskaya and etc. For pumps with a low inertia of moving masses with sufficient accuracy, it can be used the "method of intersecting characteristics", proposed by V.S. Dikarevskim [4, 5]. As a result, by using the solution of characteristics, the source of perturbation of the flow, the characteristics equation of unsteady fluid motion, it can determine the pressure in the hydraulic shock caused by the quick opening valve or pump start-up. The greatest development in the theory of water hammer was analytical methods of calculation. L. Allievi [15, 18] investigated the hydraulic shock in a simple pipeline (i.e. having a constant diameter and constant speed of propagation of shock waves), by using the general solution of differential equations of unsteady pressure flow. Zhukovsky [108] derived the equations of water hammer in finite differences, which later was called the chain of equations L.A. Allievi [18], which were subsequently was used by many researchers in the calculation of water hammer. Using the "method of characteristics" at computer simulation for transients in pressure systems was showed by K.P. Wisniewski [3, 127]. In that works, water hammer was determined by the interaction between the pressure waves that occurred at the pump and reflected in the pipeline. Loss of pressure happened conditionally apart along the pipeline. This method also allows choosing the number and size of shockproof. Development of algorithms for software simulation of transients by K.P. Vishnevsky [2, 3] was made for the complex pressure systems. It included the possible formation of discontinuities flows, hydraulic resistance, structural features of the pumping of water systems (pumps, piping, valves, etc.). However, a calculation of water hammer is adapted to high-pressure water systems for household and drinking purposes. K.P. Vishnevsky used "characteristics method" for the calculation of water hammer on a computer dedicated to the work of B.F. Lyamaeva [7]. They described in detail the process of modeling the unsteady fluid flow in complex piping systems transporting drinking water. Their works were included the description of this phenomenon at discontinuities flow, unsteady friction, changes in gas content and other parameters. Much attention was paid to the way that the original data using a grid, allowing the easiest way to record all raw data for all sites for example, large and extensive network. B.F. Lyamaevym [7, 9] developed a "method of characteristics" for the calculation of water hammer by using computer technology. The process of entering basic information for the calculation of water hammer is simplified for the user by the use of geo-information systems. Therefore software can be carried out for multiple calculations of unsteady flow regimes (pressurized systems) in transporting uncontaminated water. Calculations of hydraulic shock in multiphase systems, including a computer, are devoted to the work of V.M. Alysheva [1]. In that work, integration of differential equations of unsteady pressure flow is also performed by the "method of characteristics". The works of Streater, [148] K.P. Vishnevsky [3], B.F. Lyamaeva [7], V.M. Alyshev [1] use the method of calculation of water hammer. They are based on replacing the distributed along the length of the flow of gas parameters concentrated in the fictitious air-hydraulic caps installed on the boundaries of the pipeline. A fictitious elastic element is replaced by elastic deformation of the pipe walls, and the elastic deformation of the solid suspension is modeled by fictitious elastic elements of the solid suspension. However, detailed experimental studies are based on the solid component. A particular challenge in terms of calculations is a hydraulic shock, accompanied by discontinuities of the flow. This phenomenon has not been fully studied, so the works of many scientists are devoted to experimental and theoretical studies of non-stationary processes with a break of the column of liquid. First detailed study and writing the first design formula, for such cases of hydraulic shocks with discontinuities of continuous flow, was the work of A.F. Masty. Subsequently, the development of this issue was paid more attention by A. Surin, L. Bergeron, L.F. Moshnin, N.A. Kartvelishvili, M. Andriashev, V.S. Dikarevsky, K.P. Wisniewski, B.F. Laman, V.I. Blokhin, L.S. Gerashchenko, V.N. Kovalenko, and others. The most detailed experimental and theoretical study of water hammer with a discontinuity in the flow conduits performed by D.N. Smirnov and L.B. Zubov. As a result of the research, they describe the basic laws of gap columns, fluid and obtained relatively simple calculation dependences. In the above works, there are methods of determining maximal pressures after the discontinuities of the flow. However, the results of calculations by these methods are often contradictory. In addition, not clarified the conditions under which the maximum pressure generated. There is little influence of loss of pressure, vacuum, nature and duration of flow control and other factors on the value of maximum pressure [143,144]. The study of V.S. Dikarevsky for water hammer was included to break the continuity of flow. His work dealt with in detail, the impact magnitude of the vacuum on the course of the entire process of water hammer. Analytically and based on experimental data, scholars have argued that in a horizontal pipe rupture. The continuity of the flow occurs mainly in the regulatory body, and cavitation phenomena on the length of the pipeline are manifested. It investigates only in the form of small bubbles, whose influence on the process of hydraulic impact is negligible. As a result, research scientists have obtained analytic expressions for the hydraulic shock. They mention a gap of continuous flow, taking into account the energy loss, while controlling the flow and the wave nature. However, studies of V.S. Dikarevskogo were conducted mainly for the horizontal pressure pipelines and pumping units with a low inertia of moving masses. Researches of N.I. Kolotilo [6] and others devoted to the study of water hammer to break the continuity of flow in the intermediate point. N.I. Kolotilo analytically derived the condition for the gap of continuous flow at a turning point of the pipeline when the pressure is reduced at this point (below atmospheric pressure). Studies have shown that the location of the discontinuity of continuous flow at a turning point depends, first of all, the profile of the pipeline.

Protection of hydraulic systems against water hammer by releasing part of the transported fluid is the most widespread method of artificial reduction of the hydraulic shock. Devices that perform this function can be divided into valve, bursting disc and the overflow of the column. Membranes, as a single unit of action, which when triggered would be emptied all the contents of pipes, have not found application in dewatering. Overflow column due to high pressure and large geodetic height of these systems also do not apply. Valve device for protection against water hammer can be divided into safety valves and special shock absorbers. Safety valves in all types have a number of specific deficiencies. There is a big pressure difference between the opening and closing valves, different slamming the gate and generate an additional blow to the moment approach of negative pressure wave and the consequent need for test positives [150].

A persistence setting of safety valves prevents quenching pressure shocks (starting with lowering the pressure). Because they do not react to the effect of lowering the pressure in the pipelines at the time of the negative wave. All this has resulted in dewatering facilities are not adopted, however, as in other mining hydraulic systems [168]. Application of air hydraulic valves as a means to combat water hammer is widely used in hydraulic engineering (Especially in the systems of hydraulic transport).The valve device may not operate reliably due to the presence in the flow of solids. The valves are easy to design and necessary (in accordance volume of compressed air hydraulic parameters) and protection against pressure fluctuations, including high.

In practice, there are other methods of protection against water hammer as well as air-inlet and a fluid drain through the pump, check valves of the pipeline section, the application of equalization tanks, pressure vessel, surge tank, interference suppressors, the introduction into the pipeline of elastic elements, etc. [145,146].

1 Fluid Interpenetration and Pulsation

CONTENTS

1.1 INTRODUCTION

Mixtures of liquids, which are formed due to deep penetration (mixing) in the flow of two fluids between parallel plates or turbulent mixing in the pipes, are prevalent in our everyday lives. Deep penetration of the mixture components can be considered as a combination of diffusion and physical mixing. The first process dominates in the deep penetration and the second stage the liquids are mixed. Immiscible mixing is also crucial in modern technology. It allows chemists to control chemical reactions for the production of polymeric materials with unique properties and distribution of additives that reduce the viscous friction in the pipes. However, despite its popularity, both in nature and in the production, the mixing process is still not completely clear. Researchers in different areas can not yet even set a common terminology for it by using different names. The mixing process is extremely complex and is found in a variety of systems. Theory of mixing is included, for example soluble and partially soluble, chemically active and inert liquid, slow laminar flows, and rapid turbulent flows. Not surprisingly, there is no single theory capable to explain in detail the process of mixing in fluids. Therefore, the direct calculation is impossible to cover all important aspects of this phenomenon. Nevertheless, some information about the process of mixing can be obtained both through physical experiments and using computer simulation. Typically, in certain places or local points of water pipelines exists sharp changes in pressure above atmospheric pressure. If there is a leak in the pipeline or through the valves, spaces in the plumbing lines, is filled with gaseous phase (e.g., steam at ambient temperature) or air. The complex microscopic interaction between the components of liquid-gas mixture makes the simulation extremely important. There have been several special studies by using computer graphics in the light of immiscible mixtures. However, there are few works that deal with blending fluid. Changes in the properties of liquids in pipes and channels are due to factors such as decompression (because of the sudden opening of the discharge valve), the spread of the pulse pressure, heating or cooling, or energy production systems, mixing with the particulate matter or other body fluids (which can change the density of the liquid, specific gravity, and viscosity), the formation and collapse of vapor bubbles (cavitations) and air leak or disconnection of the system (near the vent and/or pressure wave). Changes to the boundaries of the system are due to factors such as the rapid opening or closing a valve, pipe explosion (due to high pressure) or the collapse of the tube (due to low pressure), stop the pump inlet air in the vacuum circuit breaker, the penetration of water through the valve, massive outflow valve in the discharge pressure or fire hoses, damage to the disk and/or resonance in the switching valve. Such sudden changes make the transition pressure pulse, which quickly spreads far from the place of origin of perturbations, in any possible direction, and across the sealed system. Most of the transients in water and drainage systems are the result of changes in the boundaries of the system. It happens usually near the end of the system upstream or downstream or in the local high points. Consequently, the results will help reduce the risk of system damage or failure with the proper analysis to provide a dynamic response to the shortcomings of the system, design protection equipment to manage the transition energy and determine the operational procedures to avoid transients. Analysis, design and operating procedures, that is all the benefits of computer simulations. Study of hydraulic transients began with the work of Zhukovsky [1] and Allievi [2]. Many researchers have made significant contributions in this area, including Wood [3], Angus and Parmakian [4-5], who popularized and perfected the graphical method of calculation. Benjamin Wylie and Victor Streeter [6-11], method of characteristics combined with computer modeling. Subject of transients in liquids are still growing fast around the world. Brunone et. al [12], Koelle and Luvizotto [13], Filion and Karney [14], Hamam and McCorquodale [15], Savic and Walters [16-17], Walski and Lutes [18], Wu and Simpson [19], have been developed various methods of investigation of transient pipe flow. These ranges of methods are included by approximate equations to numerical solutions of the nonlinear Navier–Stokes equations. Basic theory of unsteady fluid flow in pressure pipelines were set out in the works of Zhukovsky. He obtained the differential equations of motion of inviscid fluid formed the basis development of the theory about pressure and pressure flow of viscous fluid. With the help of this theory, it became to explanation of the physical phenomenon, known as water hammer. N. E. Zhukovsky introduced the concept of the effective sound speed. He mentioned to reducing the motion of a compressible fluid in an elastic cylindrical pipe to the motion of a compressible fluid in a rigid pipe, but with a lower modulus of elasticity of the liquid. Further study of transients in the pipelines pursuits in the works of I. A. Charny, Khristianovich, A. H. Mirzajanzadeh, M. A. Hussein-Zadeh, V. A. Yufina, H. N. Nizamova, R. F. Ganieva, L. B. Kublanovskaya, L. Polyansky, A.K. Galliamova, M.V. Lur'e, E.V. Vyazunova, A.G. Gumerova, A. Shumaylova, A. Kozak, A. A. Kandaurova, E. M. Klimovskaya, and so on. For pumps with a low inertia of moving masses with sufficient accuracy, it can be used the "method of intersecting characteristics", proposed by

Dikarevskim [20-21]. As a result, by using the solution of characteristics, the source of perturbation of the flow, the characteristics equation of unsteady fluid motion, it can determine the pressure in the hydraulic shock caused by the quick opening valve or pump start-up. The greatest development in the theory of water hammer was analytical methods of calculation. Allievi [22-23] investigated the hydraulic shock in a simple pipeline (i.e. having a constant diameter and constant speed of propagation of shock waves), by using the general solution of differential equations of unsteady pressure flow. Zhukovsky [1] derived the equations of water hammer in finite differences, which later was called the chain of equations Allievi [2], which were subsequently was used by many researchers in the calculation of water hammer. Using the "method of characteristics" at computer simulation for transients in pressure systems was showed by Wisniewski [25-26]. In that works, water hammer was determined by the interaction between the pressure waves that occurred at the pump and reflected in the pipeline. Loss of pressure happened conditionally apart along the pipeline. This method also allows choosing the number and size of shockproof. Development of algorithms for software simulation of transients by Vishnevsky [25, 27] was made for the complex pressure systems. It included the possible formation of discontinuities flows, hydraulic resistance, structural features of the pumping of water systems (pumps, piping, valves, etc.). However, a calculation of water hammer is adapted to high-pressure water systems for household and drinking purposes. K. P. Vishnevsky used "characteristics method" for the calculation of water hammer on a computer dedicated to the work of Lyamaeva [28]. They described in detail the process of modeling the unsteady fluid flow in complex piping systems transporting drinking water. Their works were included the description of this phenomenon at discontinuities flow, unsteady friction, changes in gas content and other parameters. Much attention was paid to the way that the original data using a grid, allowing the easiest way to record all raw data for all sites for example, large and extensive network. Lyamaevym [28-29] developed a "method of characteristics" for the calculation of water hammer by using computer technology. The process of entering basic information for the calculation of water hammer is simplified for the user by the use of geo-information systems. Therefore software can be carried out for multiple calculations of unsteady flow regimes (pressurized systems) in transporting uncontaminated water. Calculations of hydraulic shock in multiphase systems, including a computer, are devoted to the work of Alysheva [30]. In that work, integration of differential equations of unsteady pressure flow is also performed by the "method of characteristics". The works of Streater [31], Vishnevsky [25], Lyamaeva [28], Alyshev [30] use the method of calculation of water hammer. They are based on replacing the distributed along the length of the flow of gas parameters concentrated in the fictitious air-hydraulic caps installed on the boundaries of the pipeline. A fictitious elastic element is replaced by elastic deformation of the pipe walls, and the elastic deformation of the solid suspension is modeled by fictitious elastic elements of the solid suspension. However, detailed experimental studies are based on the solid component. A particular challenge in terms of calculations is a hydraulic shock, accompanied by discontinuities of the flow. This phenomenon has not been fully studied, so the works of many scientists are devoted to experimental and theoretical studies of non-stationary processes with a break of the column of liquid.

First detailed study and writing the first design formula, for such cases of hydraulic shocks with discontinuities of continuous flow, was the work of A. F. Masty. Subsequently, the development of this issue was paid more attention by A. Surin, L. Bergeron, L. F. Moshnin, N. A. Kartvelishvili, M. Andriashev, V. S. Dikarevsky, K. P. Wisniewski, B. F. Laman, V. I. Blokhin, L. S. Gerashchenko, V. N. Kovalenko, and others.

The most detailed experimental and theoretical study of water hammer with a discontinuity in the flow conduits performed by D. N. Smirnov and L. B. Zubov. As a result of the research, they describe the basic laws of gap columns, fluid and obtained relatively simple calculation dependences. There are methods of determining maximal pressures after the discontinuities of the flow. However, the results of calculations by these methods are often contradictory. In addition, not clarified the conditions under which the maximum pressure generated. There is little influence of loss of pressure, vacuum, nature, and duration of flow control and other factors on the value of maximum pressure [32-33]. The study of V. S. Dikarevsky for water hammer was included to break the continuity of flow. His work dealt with in detail, the impact magnitude of the vacuum on the course of the entire process of water hammer. Analytically and based on experimental data, scholars have argued that in a horizontal pipe rupture. The continuity of the flow occurs mainly in the regulatory body, and cavitations phenomena on the length of the pipeline are manifested. It investigates only in the form of small bubbles, whose influence on the process of hydraulic impact is negligible. As a result, research scientists have obtained analytic expressions for the hydraulic shock. They mention a gap of continuous flow, taking into account the energy loss, while controlling the flow and the wave nature. However, studies of V. S. Dikarevskogo were conducted mainly for the horizontal pressure pipelines and pumping units with a low inertia of moving masses. Researches of Kolotilo [34] and others devoted to the study of water hammer to break the continuity of flow in the intermediate point. N. I. Kolotilo analytically derived the condition for the gap of continuous flow at a turning point of the pipeline when the pressure is reduced at this point (below atmospheric pressure). Studies have shown that the location of the discontinuity of continuous flow at a turning point depends, first of all, the profile of the pipeline.

Protection of hydraulic systems against water hammer by releasing part of the transported fluid is the most widespread method of artificial reduction of the hydraulic shock. Devices that perform this function can be divided into valve, bursting disc and the overflow of the column. Membranes, as a single unit of action, which when triggered would be emptied all the contents of pipes, have not found application in dewatering. Overflow column due to high pressure and large geodetic height of these systems also do not apply. Valve device for protection against water hammer can be divided into safety valves and special shock absorbers. Safety valves in all types have a number of specific deficiencies. There is a big pressure difference between the opening and closing valves, different slamming the gate and generate an additional blow to the moment approach of negative pressure wave and the consequent need for test positives [35].

A persistence setting of safety valves prevents quenching pressure shocks (starting with lowering the pressure). Because they do not react to the effect of lowering the pressure in the pipelines at the time of the negative wave. All this has resulted in dewatering facilities are not adopted, however, as in other mining hydraulic systems [36]. Application of air hydraulic valves as a means to combat water hammer is widely used in hydraulic engineering (Especially in the systems of hydraulic transport). The valve device may not operate reliably due to the presence in the flow of solids. The valves are easy to design and necessary (in accordance volume of compressed air hydraulic parameters) and protection against pressure fluctuations, including high.

In practice, there are other methods of protection against water hammer as well as air-inlet and a fluid drain through the pump, check valves of the pipeline section, the application of equalization tanks, pressure vessel, surge tank, interference suppressors, the introduction into the pipeline of elastic elements, and so on. [37-38].

1.2 EMERGENCE AND PROPAGATION OF WATER HAMMER IN THE WATER PIPELINE

Water hammer is the result of sharp changes of fluid pressure by the instantaneous changes in the rate of flow in the pipeline [39-41]. This phenomena occurs during water hammer are explained on the basis of compressibility of liquid drops. After closing the valves on the horizontal pipe of constant diameter, which moves the liquid with an average speed V_0, a liquid layer, located directly at the gate, immediately stops [42,43]. Then successively terminate movement of the liquid layers (turbulence, counter flows) to increase with time away from the gate. It is compacted before stopping the mass of liquid. As a result of increasing pressure somewhat expanded pipe. In the tube includes an additional volume of liquid. Since the fluid is compressible, the whole of its mass in the pipeline does not stop immediately. It moves from the gate along the pipeline with some velocity c, called the speed of propagation of the pressure wave [44-45].

If the pressure at the inlet of the pipe and along its length is equal to p_0, then slugging pressure undergoes a sharp increase:

$$\Delta p_{y\partial} : p = p_0 + \Delta p_{y\partial}.$$

The Zhukovsky formula is as flowing:

$$\Delta p_{y\partial} = (C.\Delta v / g) \tag{1.1.1}$$

where g = acceleration of free fall. The speed of the shock wave is calculated by the formula:

$$= \sqrt{\frac{g.\dfrac{E_{\text{æc}}}{\rho}}{1 + \dfrac{d}{\delta}.\dfrac{E_{\text{æc}}}{E}}} \tag{1.1.2}$$

where $E_{\text{æ}}$ = modulus of elasticity of the liquid (water)

$$E_{\text{æ}} = 2.10^9 (Pa), \left(kg / m^2\right),$$

E = Modulus of elasticity for pipeline material Steel

$$E = 10^{11}(Pa),(kg/m^2),$$

d= Outer diameter of the pipe (mm),

$$\rho = \text{Density}\left(kg/m^3\right), \delta = \text{wall thickness } (mm)$$

Stopping of a second layer of liquid exerts pressure on the following layers gradually caused high pressure. It acts directly at the valve extends to the rest of the pipeline against fluid flow speed C. If the pressure at the beginning of the pipeline remains unchanged then after the shock of the initial section of the tube, it begins the reverse movement of the shock wave with the same velocity C.

At the same time in the pipe occurs fluid motion in the direction of the initial section. Upon reaching the shock wave section at the pressure valve is reduced and becomes smaller than the initial pressure before impact. Then it starts moving shock wave. Pressure wave drop happens in the direction of the start of the pipeline [46]. Cycles of increases and decreases in pressure will continue. It is iterated at intervals equal to time for dual-path shock wave length of the pipeline from the valve prior to the pipeline. Thus, the hydraulic impact of the liquid in the pipeline will perform oscillatory motion. The cause of oscillatory motion is hydraulic resistance and viscosity. It absorbs the initial energy of the liquid for overcoming the friction and is damped. Water hammer is manifested in hydro-machines various purposes. In most cases this is undesirable, leading to the destruction of pipelines.

To get a clear picture of the emergence and spread of water hammer as a field test model was chosen (Figure 1).

Measuring output for field tests and laboratory experiments were performed at 0:00 hours on 02/10/2007–02/05/2009. Field and laboratory tests were carried out the project. Pipeline connects the two reservoirs. Water is pumped from the basin for water purification. This reservoir is connected to the suction pipe (suction header) at pumping station to top original device. The pumping station was equipped with five centrifugal pumps. All these parts were connected with the existing water networks. By acceleration and deceleration process for deep penetration of the liquid and, consequently, the hydraulic transients can lead to the following physical phenomena. High or low values of pressure can be arising directly in the pipeline. They often alternated with a high to a low value and vice versa. The increase in pressure is due to a collapse of vapor cavities and is a consequence of pulsation and cavitations in the pump and piping.

The pump is activated due to pressure corresponding to the level of hydraulic pressure (EGD). It accelerates the flow, while the energy inflow to the reservoir reaches a dynamic equilibrium with friction under steady hydraulic head (UGN). At suddenly, turn off the power supply, the pump stops as a supplier hydraulic energy. So there is a rapid decrease in the (EHD) pump pulse and low pressure extends downstream to the lower basin. The low atmospheric pressure may occur in locally high point (the minimum transition structure).

The lower basin maintains the pressure at the static level of the liquid. This condition is a division of a water column. This is the interpenetration (mixing) of two liquids (two stream water mixtures) during the movement of the pressure wave in the direction of the basin. The pressure wave goes to downstream and returns to the pumping station upstream. Pressure pulses reflected to the pump, but is faced with a closed shut-off valve (designed to protect the pump against high pressure). It is reflected by the momentum of the high value of pressure to the pool (the maximum transition profile). Monotone structure "maximum transition profile" and the "minimum of the transition profile" in the water column provide the separation column passes into the pulsating structure. Friction ultimately reduces the transition energy and the system reaches the final steady state [46]. There is a static (EHD), in this case, since the pumping stops and the flow returns into the basin. Used the computer simulation techniques, which is the ideal tool to track the values of momentum, inertia and friction. This is necessary for a correct assessment of changing due to transient values of the mass and energy at the boundaries of the system. Transients are distributed throughout a sealed system (Figure 1).

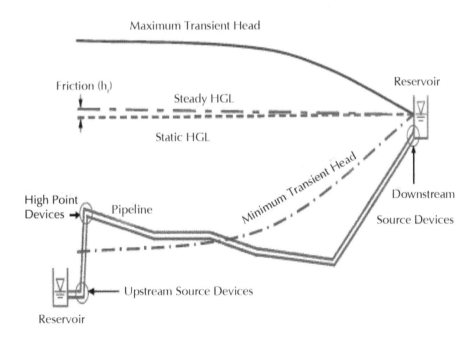

FIGURE 1 Typical locations where transient pulses initiate.

Typically, the pressure pulsations of liquid are due to the pump, which causes the acceleration and deceleration of the fluid flow [47]. This causes uncontrollable energy that manifests itself in the form of emissions of pressure [48-52].

1.3   MECHANISM OF MIXING OF TWO STREAMS IN THE WATER PIPELINE

Mechanism of fluids mixing in the water pipeline and simulation was studied by the mixing of two liquids at the experimental setup, shown in Figure 2. The cases of mutual penetration of the mixing liquids, when they are moving in separated tubes, toward a unit junction in a common stack.

All conditions for liquids, for example: speed, pressure, temperature, and other properties in tubes are identical. The challenge is to study on changes in the behavior and state of the flow of liquids. To do this, consider a model of two component fluid: liquid-liquid.

The experiments were handled by three values of velocity corresponding to Reynolds numbers of 100, 200, and 400, respectively.

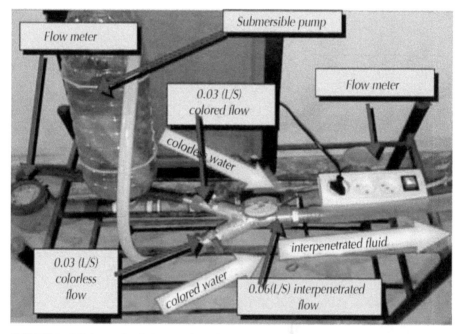

FIGURE 2   Mixing and turbulent motion of two liquids in a pipe.

To track the trajectory of the scattered fluid elements (particles) in a mixture, first liquid was colored and the second was colorless. After mixing at the joint both liquid changes color (Figure 2).

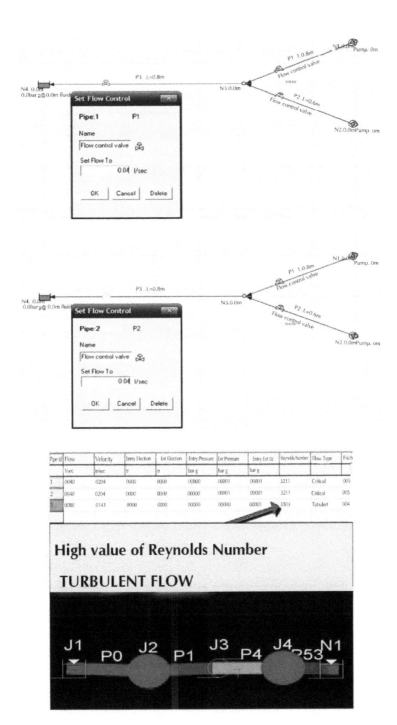

FIGURE 3 Results of laboratory experiments for the mixing of two liquids.

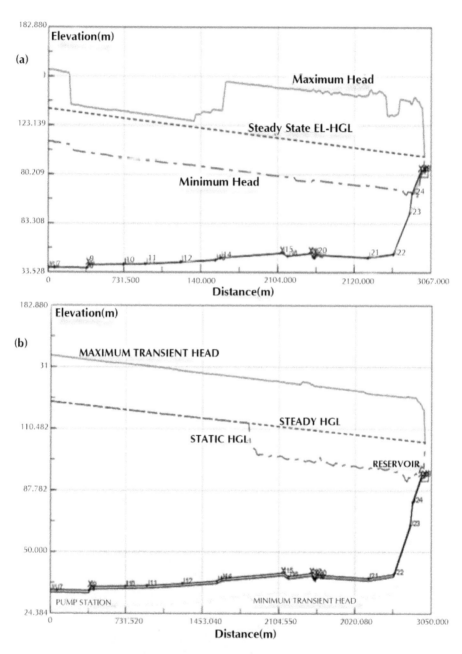

FIGURE 4   Water column separation, because of the pumps turning off.

Experiments to determine the Reynolds value when the flow becomes turbulent flow regime (Figure 4).

Water pipeline conventional pressure sensors were established in order to measuring data parameter. Fast sensors and equipments were selected for recording data in order to accurately track transient. They were connected to the response of the device to receive and collect data. The converter is connected to the pipeline system through a device that produces air.

This is necessary due to the fact that the air pressure can distort the signal transmitted during the transition process. Recording does not start until all the air is not released from the pipeline.

The time of registration data from all devices and equipments are synchronized. In protective devices, such as switching tanks (hydro-pneumatic or surge tanks), water level measured over a long time.

In this case study hydraulic shock was analyzed in four conditions:

First: There was a leak from the water pipeline.

Second: There was no leakage from the water pipeline.

Third: There was a leak in the pipeline in the presence of a surge tank or the pulse of the tank.

Fourth: There was no leakage from the pipeline in the presence of a surge tank or the pulse of the tank.

Figure 4 shows the results of field tests in the water pipeline, which was showed in Figure 1.

In this case study separation of the water column in the water pipeline happened due to pump shutdown:

(a) With the surge tank–if the pipeline leaks,

(b) Without surge tank–in the absence of leakage of the pipeline.

1.4 DETERMINATION OF FUNCTIONAL DEPENDENCE BETWEEN THE PHYSICAL PARAMETERS

1.4.1 Regression Analysis

The pipeline system has a rather complicated structure and consists of structural sections, simple piping and structural components. These devices violate the homogeneity of the pipeline (for example, these include a change of transverse size of pipes, branching, sharp turns, place the installation of pumps, valves, accumulators, etc.). In most cases, there are no probabilities to build mathematical models that adequately describe the processes occurring in water systems at the real world. Therefore, we formulate a slightly different approach to solve this problem. The essence of the approach is that seeks to determine the functional relationship between the physical units for hydraulic impact between the pressure p and different variables such as density ρ, the wave velocity C, the acceleration due to gravity g, water flow (discharge) V, pipe diameter d, pipe modulus of elasticity E_p, modulus of elasticity for water E_w, time T, are as the data of field tests.

The regression relationships were used for measured data based on specific pipe-line system (Figure 1). Various types of equations describe the structure of the interpenetration of fluids and hydraulic impact. They are applied computer software statistical support for identifying of regression equations coefficients.

To construct models for determining the speed of the shock or acoustic waves, there were two assumptions:

Assumption I: $p = f(V)$

The results of calculations are shown in the Table 1 and in Figure 5 and Figure 6.

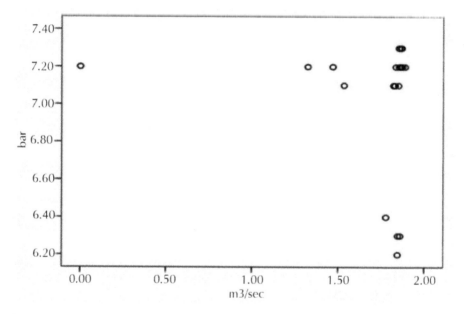

FIGURE 5　Scatter plot data and field trials p and V.

Calculations show that the regression cannot be used logarithmic and power model.

Assumption II: $p = f(V, L, T)$

To calculate the coefficients of the statistical model conducted 625 experiments on the parametric model. The calculation results are shown below in the Table 1 and in Figure 6 and Figure 7.

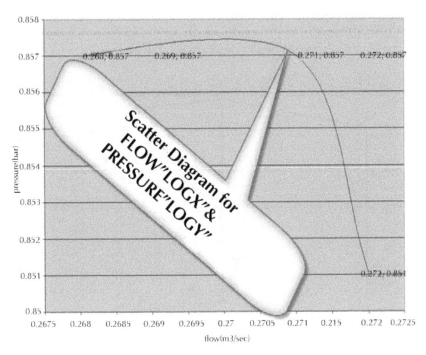

FIGURE 6 Simulation of field trials for the surge wave in hydraulic impact.

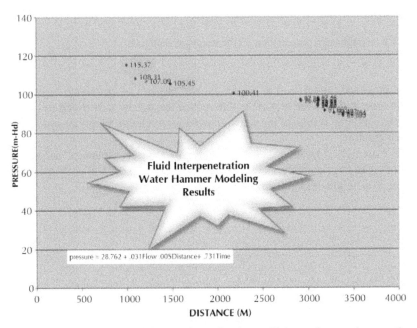

FIGURE 7 The results of calculations to determine the coefficients of regression equations.

For the water pipeline under consideration we obtain the following relation-ship.

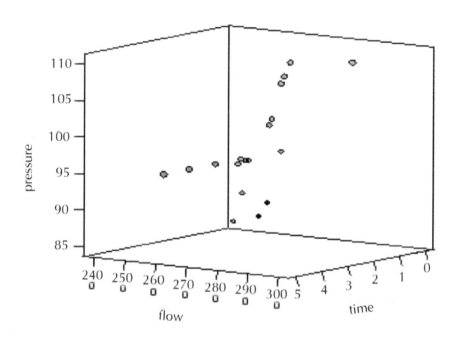

FIGURE 8   Graphical representation for the results of field tests.

The proposed model used in the design and conduct of examination parts of the pipeline leak. It allows the best method to reduce leakage by identifying its location.

Implementation results of the study allowed for a reliable supply of water through the main pipeline.

It is extremely relevant to identify a system of extreme transient pressures (for long pipes with large diameters and the mass of water). The existent profiles have large differences in height, and initial rate exceeding $1\,(m)$.

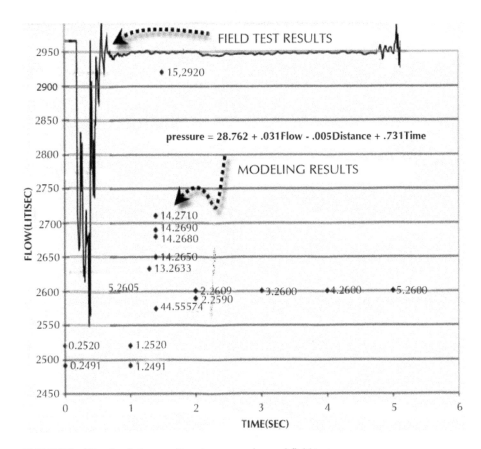

FIGURE 9   The simulation results using regression and field tests.

Figure 9 shows a comparison of simulation results using regression and field tests. The achieved results are close enough together that demonstrates the effectiveness of the proposed method.

1.5   MIXING OF LIQUIDS IN PIPE NETWORK

It is assumed that mixing of liquid in pipe network.

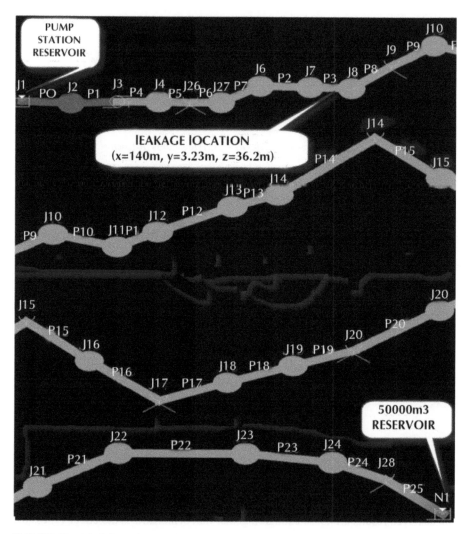

FIGURE 10   Modeling of water pipeline.

a. The independent variable $\left(\frac{m}{s}\right)$ contains non-positive values. The minimum value is .00. The logarithmic and power models cannot be calculated.

b. The independent variable $\left(\frac{m}{s}\right)$ contains values of zero. The Inverse and S models cannot be calculated. Regression equation defined in stages (2-3-7-8) has been meaningless and stages (1-4-5-6-9-10-11) have been accepted, because its coefficients are meaningful.

Transient's pressure in a single point of the system can affect all other parts of the system (Figure 10).

Thus, we obtain the following model to approximate the variables:

TABLE 1 Models and results of calculations.

Equation	Model Summary					Parameter Estimates			
	R Square	F	df1	df2	Sig.	a_0	a_1	a_2	a_3
Linear $y = a_0 + a_1 x$	0.418	15.831	1	22	0.001	6.062	0.571		
Logarithmic(a)	.	.	.	.	.	.	.		
Inverse(b)	.	.	.	.	.	.	.		
Quadratic $y = a_0 + a_1 x + a_2 x^2$	0.487	9.955	2	21	0.001	6.216	−0.365	0.468	
Cubic $y = a_0 + a_1 x + a_2 x^2 + a_3 x^3$	0.493	10.193	2	21	0.001	6.239	0	−0.057	0.174
Compound $A = Ce^{kt}$	0.424	16.207	1	22	0.001	6.076	1.089		
Power(a)	.	.	.	.	.	.	.		
S(b)	.	.	.	.	.	.	.		
Growth $(dA/dT) = KA$	0.424	16.207	1	22	0.001	1.804	0.085		

TABLE 1 *(Continued)*

Equation	Model Summary					Parameter Estimates			
	R Square	F	df1	df2	Sig.	a_0	a_1	a_2	a_3
Exponential $y = ab^x + g$	0.424	16.207	1	22	0.001	6.076	0.085		
Logistic $y = ax^b + g$	0.424	16.207	1	22	0.001	0.165	0.018		

Linear function ∴ **pressure = 6.062 + .571Flow,**
Quadratic function ∴ **pressure = 6.216 - .365Flow^4 + .468Flow^3,**
Cubic function ∴ **pressure = 6.239–.057Flow^2 + .174Flow,**
Compound function ∴ **pressure = 1.089(1+Flow)^n, n = compounding period**
Growth function ∴ **pressure = 1.804(.085)$^{Flow/.05}$,**   (1.4.1–1.4.7)
Exponential function ∴ **pressure = 6.076e$^{FlowLn.085}$,**
Logitic function ∴ **pressure = 1/(1+e^{-Flow}) or pressure = .165 + .918Flow**

Therefore, it is necessary to use a computer model to consider the specifics of each pipe. Therefore, the engineer overseeing the work of the pipeline should take into account the physical principles that govern the behavior of the network and provide a common conceptual approach to decision making. Tables 1 to Table 4 present the results of calculations using the above proposed models [53-108].

TABLE 2   Models and calculation results (Subject to hydraulic impact).

Model		Un-standardized Coefficients		Standardized Coefficients	T	Sig.
		B	Std. Error	Beta		
1	(Constant)	28.762	29.73		0.967	
	Flow	0.031	0.01	0.399	2.944	
	Distance	−0.005	0.001	−0.588	−4.356	
	Time	0.731	0.464	0.117	1.574	
2	(Constant)	14.265	29.344		0.486	
	Flow	0.036	0.01	0.469	3.533	
	Distance	−0.004	0.001	−0.52	−3.918	
3	(Constant)	97.523	1.519		64.189	
4	(Constant)	117.759	2.114		55.697	
	Distance	−0.008	0.001	−0.913	−10.033	
5	(Constant)	14.265	29.344		0.486	
	Flow	0.036	0.01	0.469	3.533	
	Distance	−0.004	0.001	−0.52	−3.918	

Accepted model of the regression is equation (1), because its prediction rates are the highest:

$$\text{pressure} = 28.762 + .031\text{Flow} - .005\text{Distane} + .731\text{Time} \qquad (1.2.8)$$

TABLE 3　Regression model, the calculated parameters, and variables.

Model		Beta in	T	Sig.	Partial Correlation	Co- linearity Statistics Tolerance
2	Time	.117(a)	1.574	0.133	0.348	0.887
3	Time	.122(b)	0.552	0.587	0.122	1
	Flow	.905(b)	9.517	0	0.905	1
	Distance	−.913(b)	−10.033	0	−0.913	1
4	Time	.189(c)	2.274	0.035	0.463	0.995
	Flow	.469(c)	3.533	0.002	0.63	0.298
5	Time	.117(a)	1.574	0.133	0.348	0.887

a. Forecast in the model: (constant), the distance, the flow;
b. Forecast: (continued);
c. forecast in the model: (constant) distance;
d. Dependent variable: pressure;

TABLE 4　Models and calculated results (Subject to hydraulic impact).

Model		Sum of Squares	df	Mean Square	F	Sig.
1	Regression	972.648	3	324.216	62.223	.000(a)
	Residual	93.791	18	5.211		
	Total	1066.439	21			
2	Regression	959.744	2	479.872	85.455	.000(b)
	Residual	106.695	19	5.616		
	Total	1066.439	21			
3	Regression	0	0	0	.	.(c)
	Residual	1066.439	21	50.783		
	Total	1066.439	21			
4	Regression	889.663	1	889.663	100.655	.000(d)
	Residual	176.775	20	8.839		

TABLE 4 *(Continued)*

Model		Sum of Squares	df	Mean Square	F	Sig.
	Total	1066.439	21			
5	Regression	959.744	2	479.872	85.455	.000(b)
	Residual	106.695	19	5.616		
	Total	1066.439	21			

a. Forecast: (constant) time, distance, flow;
b. Forecast: (constant) distance, the flow;
c. Forecast: (continued);
d. Forecast: (constant) distance

KEYWORDS

- **Method of characteristics**
- **Regression equations**
- **Safety valves**
- **Water hammer**
- **Water pipeline**

REFERENCES

1. Joukowski, N. Paper to Polytechnic Soc. Moscow, spring of 1898, *English translation by Miss O. Simin. Proc. AWWA* 57–58 (1904).
2. Allievi, L. General Theory of Pressure Variation in Pipes. *Ann. D. Ing.*, 166–171 (1982).
3. Wood, F. M. *History of Water hammer*. Civil Engineering Research Report, #65, Queens University, Canada, pp. 66–70 (1970).
4. Parmakian, J. Water hammers Design Criteria. *J. of Power Div., ASCE*, 456–460 (September, 1957).
5. Parmakian, J. *Water hammer Analysis*. Dover Publications, Inc., New York, pp. 51–58 (1963).
6. Streeter, V. L. and Lai, C. Water hammer Analysis Including Fluid Friction. *Journal of Hydraulics Division, ASCE*, **88**, 79 (1962).
7. Streeter, V. L. and Wylie, E. B. *Fluid Mechanics*. McGraw-Hill Ltd., USA, pp. 492–505 (1979).
8. Streeter, V. L. and Wylie, E. B. *Fluid Mechanics*. McGraw-Hill Ltd., USA, pp. 398–420 (1981).
9. Wylie, E. B., Streeter, V. L., and Talpin, L. B. Matched impedance to control fluid transients. *Trans. ASME*, **105**(2) 219–224 (1983).
10. Wylie, E. B. and Streeter, V. L. *Fluid Transients in Systems*. Prentice-Hall, Englewood Cliffs, New Jersey, p. 4 (1993)
11. Wylie, E. B. and Streeter, V. L. *Fluid Transients*, Feb Press, Ann Arbor, MI, p. 158 (1982, corrected copy, 1983).

12. Brunone, B., Karney, B. W., Mecarelli, M., and Ferrante, M. Velocity Profiles and Unsteady Pipe Friction in Transient Flow. *Journal of Water Resources Planning and Management, ASCE,* **126**(4), 236–244 (July, 2000).
13. Koelle, E., Luvizotto, Jr. E., and Andrade, J. P. G. *Personality Investigation of Hydraulic Networks using MOC – Method of Characteristics.* Proceedings of the 7th International Conference on Pressure Surges and Fluid Transients, Harrogate Durham, United Kingdom, pp. 1–8 (1996).
14. Filion, Y. and Karney, B. W. *A Numerical Exploration of Transient Decay Mechanisms in Water Distribution Systems.* Proceedings of the ASCE Environmental Water Resources Institute Conference, American Society of Civil Engineers, Roanoke, Virginia, p. 30 (2002).
15. Hamam, M. A. and Mc Corquodale, J. A. *Transient Conditions in the Transition from Gravity to Surcharged Sewer Flow.* Canadian J. of Civil Eng., Canada, pp. 65–98 (September, 1982).
16. Savic, D. A. and Walters, G. A. *Genetic Algorithms Techniques for Calibrating Network Models.* Report No. 95/12, Centre for Systems and Control Engineering, pp. 137–146 (1995).
17. Savic, D. A. and Walters, G. A. *Genetic Algorithms Techniques for Calibrating Network Models.* University of Exeter, Exeter, United Kingdom, pp. 41–77 (1995).
18. Walski, T. M., Lutes, T. L. Hydraulic Transients Cause Low-Pressure Problems., *Journal of the American Water Works Association,* **75**(2), 58 (1994).
19. Wu, Z. Y. and Simpson, A. R. Competent genetic-evolutionary optimization of water distribution systems. *J. Comput. Civ. Eng.,* **15**(2), 89–101 (2001).
20. Gerasimov, Yu. I. *Course of physical chemistry,* Vol. 1. M. Goskhimizdat (Ed.), p. 736 (1963).
21. Dikarevsky, M. *Impact protection opositelnyh closed systems.* Kolos, Moscow, p. 80 (1981).
22. Nigmatulin, R. I., Nagiyev, F. B., and Khabeev, N. S. *Destruction and collapse of vapor bubbles and strengthening shock waves in a liquid with vapor bubbles.* Assembly. "Gas and wave dynamics," No.3, "MSU", pp. 124–129 (1979).
23. Allievi, L. General Theory of Pressure Variation in Pipes. *Ann. D. Ing.,* pp. 166–171 (1982).
24. Joukowski, N. Paper to Polytechnic Soc. Moscow, spring of 1898. *English translation by Miss O. Simin. Proc. AWWA,* 57–58 (1904).
25. Wisniewski, K. P. *Design of pumping stations closed irrigation systems: Right.* K. P. Vishnevsky and A. V. Podlasov (Eds.). Agropromizdat, Moscow, p. 93 (1990).
26. Nigmatulin, R. I., Khabeev, N. S., and Nagiyev, F. B. *Dynamics, heat and mass transfer of vapor-gas bubbles in a liquid.* Int. J. Heat Mass Transfer, Vol. 24, N6, Printed in Great Britain, pp. 1033–1044 (1981).
27. Vargaftik, N. B. *Handbook of thermo-physical properties of gases and liquids.* Pergamon Press, Oxford, p. 98 (1972).
28. Laman, B. F. *Hydro pumps and installation,* p. 278 (1988).
29. Nagiyev, F. B. and Kadyrov, B. A. Heat transfer and the dynamics of vapor bubbles in a liquid binary solution. *DAN Azerbaijani SSR,* (4) 10–13 (1986).
30. Alyshev, V. M. *Hydraulic calculations of open channels on your PC.* Part 1 Tutorial. MSUE, Moscow, p. 185 (2003).
31. Streeter, V. L. and Wylie, E. B. *Fluid Mechanics.* McGraw-Hill Ltd., USA, pp. 492–505 (1979).
32. Sharp, B. *Water hammer Problems & Solutions.* Edward Arnold Ltd., London, pp. 43–55 (1981).
33. Skousen, P. *Valve Handbook.* HAMMER Theory and Practice. McGraw Hill, New York, pp. 687–721 (1998).

34. Shaking, N. I. *Water hammer to break the continuity of the flow in pressure conduits pumping stations.* Dis. on Kharkov, p. 225 (1988).
35. Tijsseling, A. S. and Alan E Vardy. *Time scales and FSI in unsteady liquid-filled pipe flow.* pp. 5–12 (1993).
36. Wu, P. Y. and Little, W. A. Measurement of friction factor for flow of gases in very fine channels used for micro miniature, Joule Thompson refrigerators. *Cryogenics*, **24**(8) 273–277 (1983).
37. Song, C. C. et al. Transient Mixed-Flow Models for Storm Sewers. *J. of Hyd. Div.*, **109** 458–530 (November, 1983).
38. Stephenson, D. *Pipe Flow Analysis.*, Elsevier, Vol. 19, South Australia, pp. 670–788 (1984).
39. Chaudhry, M. H. *Applied Hydraulic Transients.* Van Nostrand Reinhold Co., New York, pp. 1322–1324 (1979).
40. Chaudhry, M. H. and Yevjevich, V. *Closed Conduit Flow.* Water Resources Publication, USA, pp. 255–278 (1981).
41. Chaudhry, M. H. *Applied Hydraulic Transients.* Van Nostrand Reinhold, New York, USA, pp. 165–167 (1987).
42. Kerr, S. L. Minimizing service interruptions due to transmission line failures: Discussion. *Journal of the American Water Works Association*, **41, 634**, 266–268 (July, 1949).
43. Kerr, S. L. Water hammer control. *Journal of the American Water Works Association*, **43**, 985–999 (December, 1951).
44. Apoloniusz Kodura and Katarzyna Weinerowska. *Some Aspects of Physical and Numerical Modeling of Water Hammer in Pipelines.* pp. 126–132 (2005).
45. Anuchina, N. N., Volkov, V. I., Gordeychuk, V. A., Es'kov, N. S., Ilyutina, O. S., and Kozyrev, O. M. Numerical simulations of Rayleigh-Taylor and Richtmyer-Meshkov instability using mah-3 code. *J. Comput. Appl. Math.*, **168** 11 (2004).
46. Fox, J. A. *Hydraulic Analysis of Unsteady Flow in Pipe Network.* Wiley, New York, pp. 78–89 (1977).
47. Karassik, I. J. *Pump Handbook—Third Edition.* McGraw-Hill, pp. 19–22 (2001).
48. Fok, A. Design Charts for Air Chamber on Pump Pipelines. *J. of Hyd. Div., ASCE*, 15–74 (September, 1978).
49. Fok, A., Ashamalla, A., and Aldworth, G. *Considerations in Optimizing Air Chamber for Pumping Plants.* Symposium on Fluid Transients and Acoustics in the Power Industry, San Francisco, USA, pp. 112–114 (December, 1978).
50. Fok, A. *Design Charts for Surge Tanks on Pump Discharge Lines.* BHRA 3rd Int. Conference on Pressure Surges, Bedford, England, pp. 23–34 (March, 1980).
51. Fok, A. *Water hammer and Its Protection in Pumping Systems.* Hydro technical Conference, CSCE, Edmonton, pp. 45–55 (May, 1982).
52. Fok, A. *A contribution to the Analysis of Energy Losses in Transient Pipe Flow.* PhD. Thesis, University of Ottawa, pp. 176–182 (1987).
53. Hariri Asli, K. and Nagiyev, F. B. *Water Hammer and fluid condition, Ministry of Energy*, Gilan Water and Wastewater Co., Research Week Exhibition, Tehran, Iran, pp. 132–148, http://isrc.nww.co.ir (December, 2007).
54. Hariri Asli, K. and Nagiyev, F. B. *Water Hammer analysis and formulation.* Ministry of Energy, Gilan Water and Wastewater Co., Research Week Exhibition, Tehran, Iran, pp. 111–131,http://isrc.nww.co.ir (December, 2007).
55. Hariri Asli, K. and Nagiyev, F. B. *Water Hammer and hydrodynamics instabilities, Interpenetration of two fluids at parallel between plates and turbulent moving in pipe.* Ministry of Energy, Guilan Water and Wastewater Co., Research Week Exhibition, Tehran, Iran, pp. 90–110, http://isrc.nww.co.ir (December, 2007).

56. Hariri Asli K. and Nagiyev, F. B. *Water Hammer and pump pulsation.* Ministry of Energy, Guilan Water and Wastewater Co., Research Week Exhibition, Tehran, Iran, pp. 51–72, http://isrc.nww.co.ir (December, 2007).

57. Hariri Asli, K. and Nagiyev F. B. *Reynolds number & hydrodynamics' instability.* Ministry of Energy, Guilan Water and Wastewater Co., Research Week Exhibition, Tehran, Iran, pp. 31–50, http://isrc.nww.co.ir (December, 2007).

58. Hariri Asli, K. and Nagiyev, F. B. *Water Hammer and valves.* Ministry of Energy. Guilan Water and Wastewater Co., Research Week Exhibition, Tehran, Iran, pp. 20–30, http://isrc. nww.co.ir (December, 2007).

59. Hariri Asli, K. and Nagiyev, F. B. *Interpenetration of two fluids at parallel between plates and turbulent moving in pipe.* Ministry of Energy, Guilan Water and Wastewater Co., Research Week Exhibition, Tehran, Iran, pp. 73–89, http://isrc.nww.co.ir (December, 2007).

60. Hariri Asli, K. and Nagiyev, F. B. Decreasing of Unaccounted For Water (UFW) by Geographic Information System (GIS) in Rasht urban water system, civil engineering organization of Guilan. *Technical and Art Journal* 3–7, http://www.art-of-music.net/ (2007).

61. Hariri Asli, K. *Portable Flow meter Tester Machine Apparatus.* Certificate on registration of invention, Tehran, Iran, #010757, Series a/82, pp. 1–3 (November 24, 2007)

62. Hariri Asli, K. Nagiyev, F. B., and Haghi A. K. *Interpenetration of two fluids at parallel between plates and turbulent moving in pipe.* 9th Conference on Ministry of Energetic works at research week, Tehran, Iran, pp. 73–89, http://isrc.nww.co.ir (2008).

63. Hariri Asli, K., Nagiyev, F. B., and Haghi, A. K. *Water hammer & valves.* 9th Conference on Ministry of Energetic works at research week, Tehran, Iran, pp. 20–30, http://isrc.nww. co.ir (2008).

64. Hariri Asli, K., Nagiyev, F. B., and Haghi, A. K. Water hammer and hydrodynamics instability. 9th Conference on Ministry of Energetic works at research week, Tehran, Iran, pp. 90–110, http://isrc.nww.co.ir (2008).

65. Hariri Asli, K., Nagiyev, F. B., and Haghi, A. K. *Water hammer analysis and formulation.* 9th Conference on Ministry of Energetic works at research week, Tehran, Iran, pp. 27–42, http://isrc.nww.co.ir (2008).

66. Hariri Asli, K., Nagiyev, F. B., and Haghi, A. K. *Water hammer and fluid condition.* 9th Conference on Ministry of Energetic works at research week, Tehran, Iran, pp. 27–43, http://isrc.nww.co.ir (2008).

67. Hariri Asli, K., Nagiyev, F. B., and Haghi, A. K. *Water hammer and pump pulsation.* 9th Conference on Ministry of Energetic works at research week, Tehran, Iran, pp. 27–44, http://isrc.nww.co.ir (2008).

68. Hariri Asli, K., Nagiyev, F. B., and Haghi, A. K. *Reynolds number and hydrodynamics instability.* 9th Conference on Ministry of Energetic works at research week, Tehran, Iran, pp. 27–45, http://isrc.nww.co.ir (2008).

69. Hariri Asli, K., Nagiyev, F. B., and Haghi, A. K. *Water hammer and fluid Interpenetration* 9th Conference on Ministry of Energetic works at research week, Tehran, Iran, pp. 27–47, http://isrc.nww.co.ir (2008).

70. Hariri Asli, K. *GIS and water hammer disaster at earthquake in Rasht water pipeline.* civil engineering organization of Guilan. Technical and Art Journal, pp. 14–17, http://www.art-of-music.net/ (2008).

71. Hariri Asli, K. *GIS and water hammer disaster at earthquake in Rasht water pipeline.* 3rd International Conference on Integrated Natural Disaster Management, Tehran university, ISSN: 1735-5540, 18-19 Feb., INDM, Tehran, Iran, №13, pp. 53/1–12, http://www.civilica.com/Paper-INDM03-INDM03_001.html (2008).

72. Hariri Asli, K. and Nagiyev, F. B. *Bubbles characteristics and convective effects in the binary mixtures.* Transactions issue mathematics and mechanics series of physical-technical

& mathematics science, ISSN: 0002-3108, Azerbaijan, Baku, pp. 68–74, http://www.imm. science.az/journals.html (2009).

73. Hariri Asli, K., Nagiyev, F. B., Haghi, A. K., and Aliyev, S. A. *Three-Dimensional conjugate heat transfer in porous media.* 1st Festival on Water and Wastewater Research & Technology, Tehran, Iran, pp. 26–28, http://isrc.nww.co.ir (December 12–17, 2009).

74. Hariri Asli, K., Nagiyev, F. B., Haghi, A. K., and Aliyev, S. A. *Some Aspects of Physical and Numerical Modeling of water hammer in pipelines.* 1 st Festival on Water and Wastewater Research and Technology, Tehran, Iran, pp. 26–29, http://isrc.nww.co.ir (December 12–17, 2009)

75. Hariri Asli, K., Nagiyev, F. B., Haghi, A. K., and Aliyev, S. A. *Modeling for Water Hammer due to valves: From theory to practice.* 1st Festival on Water and Wastewater Research and Technology, Tehran, Iran, pp. 26, 30, http://isrc.nww.co.ir (December 12–17, 2009).

76. Hariri Asli, K., Nagiyev, F. B., Haghi, A. K., Aliyev, S. A. *Water hammer and hydrodynamics instabilities modeling: From Theory to Practice.* 1 st Festival on Water & Wastewater Research and Technology, Tehran, Iran, pp. 26–31, http://isrc.nww.co.ir (December 12–17, 2009)

77. Hariri Asli, K., Nagiyev, F. B., Haghi, A. K., and Aliyev, S. A. *A computational approach to study fluid movement.* 1st Festival on Water and Wastewater Research and Technology, Tehran, Iran, pp. 27–32, http://isrc.nww.co.ir (December 12–17, 2009).

78. Hariri Asli, K., Nagiyev, F. B., Haghi, A. K., and Aliyev, S. A. Water Hammer Analysis: Some Computational Aspects and practical hints, 1st Festival on Water and Wastewater Research and Technology, Tehran, Iran, pp. 27–33, http://isrc.nww.co.ir (December 12–17, 2009).

79. Hariri Asli, K., Nagiyev, F. B., Haghi, A. K., and Aliyev, S. A. *Water Hammer and Fluid condition: A computational approach.* 1st Festival on Water and Wastewater Research and Technology, Tehran, Iran, pp. 27–34, http://isrc.nww.co.ir. (December 12–17, 2009).

80. Hariri Asli, K., Nagiyev, F. B., Haghi, A. K., and Aliyev, S. A. *A computational Method to Study Transient Flow in Binary Mixtures.* 1st Festival on Water and Wastewater Research and Technology, Tehran, Iran, pp. 27–35, http://isrc.nww.co.ir. (December 12–17, 2009).

81. Hariri Asli, K., Nagiyev, F. B., and Haghi, A. K. *Physical modeling of fluid movement in pipelines.* 1st Festival on Water and Wastewater Research and Technology, Tehran, Iran, pp. 27–36, http://isrc.nww.co.ir. (December 12–17, 2009).

82. Hariri Asli, K., Nagiyev, F. B., Haghi, A. K., and Aliyev, S. A. *Interpenetration of two fluids at parallel between plates and turbulent moving.* 1st Festival on Water and Wastewater Research and Technology, Tehran, Iran, pp. 27–37, http://isrc.nww.co.ir. (December 12–17, 2009).

83. Hariri Asli, K., Nagiyev, F. B., Haghi, A. K., and Aliyev, S. A. *Modeling of fluid interaction produced by water hammer.* 1st Festival on Water and Wastewater Research and Technology, Tehran, Iran, pp. 27–38, http://isrc.nww.co.ir. (December 12–17, 2009).

84. Hariri Asli, K., Nagiyev, F. B., Haghi, A. K., and Aliyev, S. A. *GIS and water hammer disaster at earthquake in Rasht pipeline.* 1st Festival on Water and Wastewater Research and Technology, Tehran, Iran, pp. 27–39, http://isrc.nww.co.ir. (December 12–17, 2009).

85. Hariri Asli, K., Nagiyev, F. B., Haghi, A. K., and Aliyev, S. A. *Interpenetration of two fluids at parallel between plates and turbulent moving.* 1st Festival on Water and Wastewater Research and Technology, Tehran, Iran, pp. 27–40, http://isrc.nww.co.ir. (December 12–17, 2009).

86. Hariri Asli, K., Nagiyev, F. B., Haghi, A. K., and Aliyev, S. A. *Water hammer and hydrodynamics' instability*. 1st Festival on Water and Wastewater Research and Technology, Tehran, Iran, pp. 27–41, http://isrc.nww.co.ir. (December 12–17, 2009).

87. Hariri Asli, K., Nagiyev, F. B., Haghi, A. K., and Aliyev, S. A. *Water hammer analysis and formulation*, 1st Festival on Water and Wastewater Research and Technology, Tehran, Iran, pp. 27–42, http://isrc.nww.co.ir. (December 12–17, 2009).

88. Hariri Asli, K., Nagiyev, F. B., Haghi, A. K., and Aliyev, S. A. *Water hammer and fluid condition*. 1st Festival on Water and Wastewater Research and Technology, Tehran, Iran, pp. 27–43, http://isrc.nww.co.ir. (December 12–17, 2009).

89. Hariri Asli, K., Nagiyev, F. B., Haghi, A. K., and Aliyev, S. A. *Water hammer and pump pulsation*. 1st Festival on Water and Wastewater Research and Technology, Tehran, Iran, pp. 27–44, http://isrc.nww.co.ir. (December 12–17, 2009).

90. Hariri Asli, K., Nagiyev, F. B., Haghi, A. K., and Aliyev, S. A. Reynolds number and hydrodynamics instabilities. 1st Festival on Water and Wastewater Research and Technology, Tehran, Iran, pp. 27–45, http://isrc.nww.co.ir. (December 12–17, 2009).

91. Hariri Asli, K., Nagiyev, F. B., Haghi, A. K., and Aliyev, S. A. *Water hammer and valves*. 1st Festival on Water and Wastewater Research and Technology, Tehran, Iran, pp. 27–46, http://isrc.nww.co.ir. (December 12–17, 2009).

92. Hariri Asli, K., Nagiyev, F. B., Haghi, A. K., Aliyev, S. A. *Water hammer and fluid Interpenetration*. 1st Festival on Water and Wastewater Research and Technology, Tehran, Iran, pp. 27–47, http://isrc.nww.co.ir. (December 12–17, 2009).

93. Hariri Asli, K., and Nagiyev, F. B. *Modeling of fluid interaction produced by water hammer*. International Journal of Chemoinformatics and Chemical Engineering, IGI, ISSN: 2155-4110, EISSN: 2155-4129, USA, pp. 29–41, http://www.igi-global.com/journals/details.asp?ID=34654 (2010).

94. Hariri Asli, K., Nagiyev, F. B., and Haghi, A. K. Water hammer and fluid condition; a computational approach. *Computational Methods in Applied Science and Engineering*. Chapter 5, Nova Science Publications, ISBN: 978-1-60876-052-7, USA, pp. 73–94, https://www.novapublishers.com/catalog/ (2010).

95. Hariri Asli, K., Nagiyev, F. B., and Haghi, A. K. Some aspects of physical and numerical modeling of water hammer in pipelines. *Computational Methods in Applied Science and Engineering*. Chapter 23, Nova Science Publications, ISBN: 978-1-60876-052-7, USA, pp. 365–387, https://www.novapublishers.com/catalog/ (2010).

96. Hariri Asli, K., Nagiyev, F. B., and Haghi, A. K. Modeling for water hammer due to valves; from theory to practice. *Computational Methods in Applied Science and Engineering*. Chapter 11, Nova Science Publications ISBN: 978-1-60876-052-7, USA, pp. 229–236, https://www.novapublishers.com/catalog/ (2010).

97. Hariri Asli, K., Nagiyev, F. B., and Haghi, A. K. A computational method to Study transient flow in binary mixtures. *Computational Methods in Applied Science and Engineering*. Chapter 13, Nova Science Publications ISBN: 978-1-60876-052-7, USA, pp. 229–236, https://www.novapublishers.com/catalog/ (2010).

98. Hariri Asli, K., Nagiyev, F. B., and Haghi, A. K. Water hammer analysis; some computational aspects and practical hints. *Computational Methods in Applied Science and Engineering*. Chapter 16, Nova Science Publications ISBN: 978-1-60876-052-7, USA, pp. 263–281,https://www.novapublishers.com/catalog/ (2010).

99. Hariri Asli, K., Nagiyev, F. B., and Haghi, A. K. Water hammer and hydrodynamics instabilities modeling. *Computational Methods in Applied Science and Engineering*. Chapter 17, From Theory to Practice, Nova Science Publications ISBN: 978-1-60876-052-7, USA, pp. 283–301,https://www.novapublishers.com/catalog/ (2010).

100. Hariri Asli, K., Nagiyev, F. B., and Haghi, A. K. A computational approach to study water hammer and pump pulsation phenomena, *Computational Methods in Applied Science and Engineering*. Chapter 22, Nova Science Publications, ISBN: 978-1-60876-052-7, USA, pp. 349–363, https://www.novapublishers.com/catalog/ (2010).

101. Hariri Asli, K., Nagiyev, F. B., and Haghi, A. K. A computational approach to study fluid movement. *Nanomaterials Yearbook - 2009, From Nanostructures, Nanomaterials and Nanotechnologies to Nanoindustry*. Chapter 16, Nova Science Publications. ISBN: 978-1-60876-451-8, USA, pp. 181–196, https://www.novapublishers.com/catalog/product_info.php?products_id=11587 (2010).

102. Hariri Asli, K., Nagiyev, F. B., and Haghi, A. K. Physical modeling of fluid movement in pipelines. *Nanomaterials Yearbook - 2009, From Nanostructures, Nanomaterials and Nanotechnologies to Nanoindustry*. Chapter 17, Nova Science Publications, USA, ISBN: 978-1-60876-451-8, USA, pp. 197–214, https://www.novapublishers.com/catalog/product_info.php?products_id=11587 (2010).

103. Hariri Asli, K., Nagiyev, F. B., and Haghi, A. K. Some Aspects of Physical and Numerical Modeling of water hammer in pipelines. *Nonlinear Dynamics an International Journal of Nonlinear Dynamics and Chaos in Engineering Systems*, ISSN: 1573-269X (electronic version) Journal no. 11071 Springer, Netherlands, ISSN: 0924-090X (print version), Springer, Heidelberg, Germany, Number, Vol 60, pp. 677–701, http://www.springerlink.com/openurl.aspgenre=article&id=doi:10.1007/s11071-009-9624-7 (2009, June 4, 2010).

104. Hariri Asli, K., Nagiyev, F. B., and Haghi, A. K. Interpenetration of two fluids at parallel between plates and turbulent moving in pipe; a case study. *Computational Methods in Applied Science and Engineering*. Chapter 7, Nova Science Publications, ISBN: 978-1-60876-052-7, USA, pp. 107–133, https://www.novapublishers.com/catalog/ (2010).

105. Hariri Asli, K., Nagiyev, F. B., Beglou, M. J., and Haghi, A. K. Kinetic analysis of convective drying. *International Journal of the Balkan Tribological Association*. ISSN: 1310-4772, Sofia, Bulgaria, **15**(4) 546–556, jbalkta@gmail.com (2009).

106. Hariri Asli, K., Nagiyev, F. B., and Haghi, A. K. Three-dimensional Conjugate Heat Transfer in Porous Media. *International Journal of the Balkan Tribological Association*, ISSN: 1310-4772, Sofia, Bulgaria, **15**(3) 336–346, jbalkta@gmail.com (2009).

107. Hariri Asli, K., Nagiyev, F. B., Haghi, A. K., and Aliyev, S. A. Pure Oxygen penetration in wastewater flow. *Recent Progress in Research in Chemistry and Chemical Engineering*. Nova Science Publications, ISBN: 978-1-61668-501-0, USA, pp. 17–27, https://www.novapublishers.com/catalog/product_info.php?products_id=13174110100 (2010).

108. Hariri Asli, K., Nagiyev, F. B., Haghi, A. K., and Aliyev, S. A. Improved modeling for prediction of water transmission failure. *Recent Progress in Research in Chemistry and Chemical Engineering*. Nova Science Publications, ISBN: 978-1-61668-501-0, USA, pp. 28–36, https://www.novapublishers.com/catalog/product_info.php?products_id=13174 (2010).

2 The Interpenetration of Fluids in the Pipe and Stratified Flow in the Channel

CONTENTS

2.1 INTRODUCTION

In the dewatering plants there is a risk of water hammer as long as increasing depth of piping and increasing the length for pipeline. The increase in flow velocities in the pipes, will be increased the value of hydraulic shock. This may lead to the destruction of the pipeline and violations of the normal operation of dewatering systems. The emergency power shutdown of engine pumps is the main reason which to cause the water hammer pressure in pipelines. A hydraulic shock arises from the changes in the degree of opening the valves. It is practically possible to avoid changing the mode of closing and opening of the valves [1, 2].

The problems of protection against water hammer solution are formed by increasing the safety margin of pipes. This can be attributed to the imperfect design of the device proposed for protecting pipelines from water hammer. It is a complex process, nonlinearly depending on factors such as the moment of inertia of the rotor pump unit. It is related to the length of pressure pipe and so on [3-4].

We well know that the plants with lower height probable with high relative increase in pressure. Therefore, it is necessary the protection of dewatering equipment necessary to apply the device for protection against water hammer.

The suggestion of the suitable device against water hammer must be recognized in the design of a new pumping installation. The value of excess pressure (working pressure of the system) in the hydraulic impact must be revealed. This problem can be solved in several ways. It usually requires a lot of time consuming and complex calculation that a person is not always a virtue. For this purpose, it has been developed abroad program for calculating water hammer in the pipeline. But as a rule, the cost of the software is sufficiently large and, therefore, design organizations and research institutions have limited ability to use software products of foreign firms [5].

The sudden opening of the valve (i.e., the rapid exchange of fluid velocity) which causes a decrease in pressure occurs in oscillatory process and pressure change. In cases with the impact phenomenon, the system must be equipped with devices that are not allowed to make an instant decrease in velocity (valve-type shut-off devices). This device shall be restricting the spread of pressure wave attack.

By Geography Information Systems (GIS), water hammer can be recognized during irregular operations. Under abnormal operations (during an earthquake) at power turning on and off, water hammer will attack to the system. The equipment, which can help during an emergency outage, includes surge tank, switching tanks, boilers, valves, pressure vessel for the pulse pressure damping, and pressure relief valves.

In the application of the coordination geometry of the pipeline, it is necessary to background information in the database which managed by the GIS. It needs to the exchange the data between the receiver and transmitter. The pipeline was equipped with the Program Logic Control (PLC). It can be controlled the entire system in online mode, by sending voltage to the valves, pumps, protective shells from PLC. In this way it can protect the system from the water hammer disaster. In some cases, hydraulic force in the transitional flow regimes lead to cracks or ruptures of pipes, even at low speeds and steady flow.

It will also explore the influence of system topology, the characteristics of fluid on the most likely causes for transients.

2.2 EQUATIONS DESCRIBING THE TRANSIENTS IN PIPELINES

An important application of the results of these studies is it is applications in the online mode control of water hammer. This methodology is based on the calculation of transients by using the automated software device. This software package allows taking into account the leakage of water in the water pipeline. It determines the leak; identify the presence of water hammer and to control fluid flow. Differential equations included in the proposed model describe the unsteady motion of a real fluid through the pipes.

These differential equations are derived from the following assumptions. It was assumed that the pipe is cylindrical with a constant cross-sectional area with the initial pressure. The fluid flow through the pipe is the one dimensional. It is assumed that the characteristics of resistors, fixed for steady flows and unsteady flows are equivalent.

Pipe wall is elastic and cross sectional area of the pipe is described by a linear dependence on pressure. The rate of fluid flow is less than the speed of sound C. Liquid and its density ρ depend linearly on pressure p. The system of governing equations written in the following form [126]:

$$(\partial v / \partial t)(1/ p)(\partial p / \partial s) + g(dz / ds) + (f / 2D)v|v| = 0, \quad \text{(Euler equation) (2.1.1)}$$

$$C^2(\partial v / \partial s) + (1/ p)(\partial P / \partial t) = 0, \quad \text{(Continuity equation) (2.1.2)}$$

The method of characteristics (MOC) was applied to solve partial differential equations.

According to the MOC we take a linear combination of Euler and continuity:

$$\lambda[(\partial v / \partial t) + (1/ p)(\partial p / \partial s) + g(dz / ds) + (f / 2D)v|v|] + C^2(\partial v / \partial s) + (1/ p)(\partial p / \partial t) = 0, \quad (2.1.3)$$
$$\lambda = {}^+ c \, \& \, \lambda = {}^- c$$

$$(dv / dt) + (1/ cp)(dp / ds) + g(dz / ds) + (f / 2D)v|v| = 0 \quad (2.1.4)$$

$$(dv / dt) - (1/ cp)(\partial p / \partial s) + g(dz / ds) + (f / 2D)v|v| = 0, \quad (2.1.5)$$

In the coordinates $((s - t))$ (Figure 1) we have:

$$(dv / dt) - (g / c)(dH / dt) = 0 \quad (2.1.6)$$

$$H_2 - H_1 = (C / g)(v_2 - v_1) = pC(v_2 - v_1), (dp = (c / g)dv,) \quad \text{(Zhukovsky Formula)} \quad (2.1.7)$$

Equations (2.1.1)–(2.1.7) cannot be solved analytically, but they may be expressed graphically in the space time as a characteristic line (or lines). It is called features, which are signals that propagate to the right (C^+) and left (C^-) simultaneously from each location in the system.

At each interior point, signals arriving from two adjacent points simultaneously. A linear combination of H and V is invariant along each characteristic [6-7]. Therefore, H and V can be obtained from the exact points [8-9].

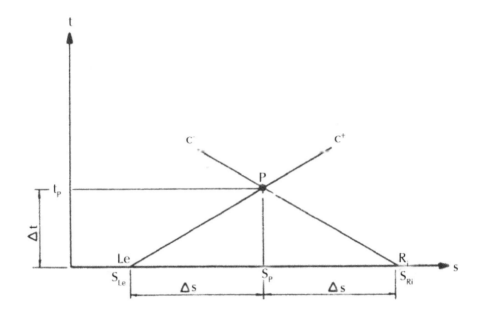

FIGURE 1   Determination of the characteristic lines intersection (ST).

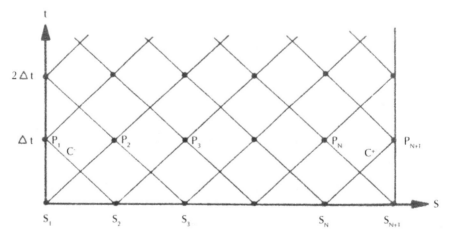

FIGURE 2   Determination of the characteristic lines coordination (ST).

It was used the finite difference method (Figure 2):

$$c+ : (vp - v_{le})(Tp - 0) + (Hp - H_{Le})/(Tp - 0) + (fv_{Le}|v_{Le}|)/2D) = 0| \qquad (2.1.8)$$

$$c- : (vp - vRi)(Tp - 0) + (g/c)(Hp - HRi)/(Tp - 0) + (fvRi|vRi|)/2D = 0| \qquad (2.1.9)$$

$$c+ : (vp - v_{Le}) + (g/c)(Hp - H_{Le}) + (f\Delta t)(fv_{Le}|v_{Le}|)/2D = 0 \qquad (2.1.10)$$

$$c- : (vp - vRi) + (g/c)(Hp - HRi) + (f\Delta t)(fv_{Ri}|v_{Ri}|)/2D = 0 \qquad (2.1.11)$$

$$V_p = 1/2 \begin{pmatrix} (V_{Le} + V_{ri}) + (g/c)(H_{Le} - H_{ri}) \\ -(f\Delta t/2D)(V_{Le}|V_{Le}| + V_{ri}|V_{ri}|) \end{pmatrix} \qquad (2.1.12)$$

$$H_p = 1/2 \begin{pmatrix} C/g(V_{Le} + V_{ri}) + (H_{Le} + H_{ri}) \\ -C/g(f\Delta t/2D)(V_{Le}|V_{Le}| - V_{ri}|V_{ri}|) \end{pmatrix} \qquad (2.1.13)$$

Thus, the pressure in the hydraulic impact or pressure wave (ΔH) is a function of the following independent variables [10]:

$$\Delta H = \Delta H(f, T, C, v, g, D) \qquad (2.1.14)$$

2.3 RESULTS OF CALCULATIONS BASED ON FIELD TEST OF THE MODEL

Figure 3 and Figure 4 show the results of calculations obtained using the software calculation of water hammer by the MOC. These graphs show the flow and pressure variations against the time.

Figure 3 shows the character of the dependence of flow to time for the water pipeline (without surge tank case). The leakage is close to harmonic vibrations. It is seen that within 1.3 s of the flow decreases from 2950 $\left(lit/_s \right)$ to a minimum of 2500 $\left(lit/_s \right)$.

Then it takes the form of harmonic oscillations. It is showed the corresponding dependence of the flow-time and pressure-time variations on surge tank by Figure 4. Also it is subjected to leakage which is shown in Figure 4. In this case, the flow is reduced from 2,950 $\left(lit/_s \right)$ to a minimum of 2,520 $\left(lit/_s \right)$ after 6 s. Then after 4 s the flow increases up to 3,000 $\left(lit/_s \right)$.

Thus, for 1 second from the surge tank, 125 l of water drained by the leakage point.

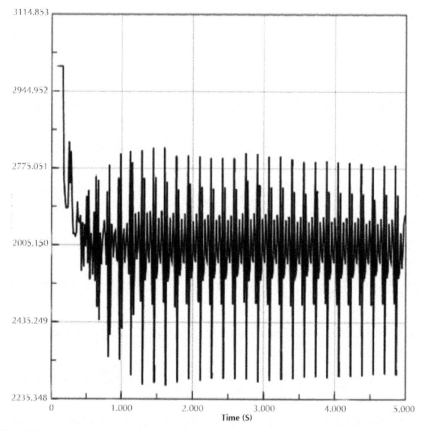

FIGURE 3 The dependence of the flow to time for transmission line without surge tank and in leakage condition.

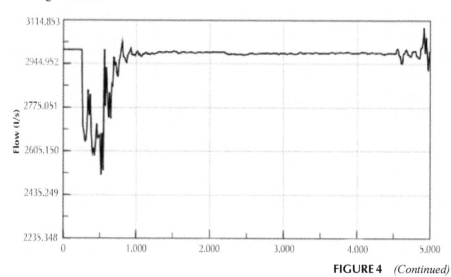

FIGURE 4 *(Continued)*

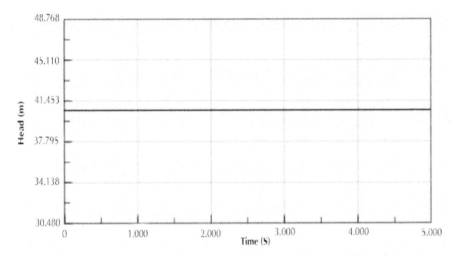

FIGURE 4   Dependence of flow and pressure to time for transmission line with surge tank and in leakage condition.

Figure 5 shows the results of calculations obtained by using the software for calculating water hammer (the method of characteristics). This Figure 5 shows the maximum pressure against the distance.

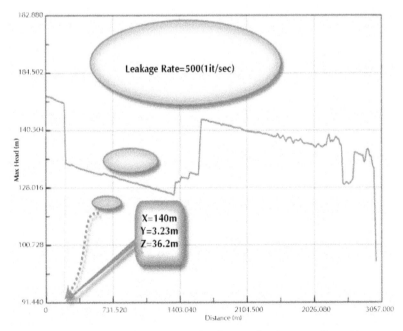

FIGURE 5   The dependence of pressure on the distance for the water line with a surge tank, subject to leakage.

Consider an elastic model, which assumes that the propagation of the surge pressure distribution in pipeline and the liquid is linearly elastic. As the liquid is not completely incompressible, during the transition of the pressure wave propagation its density may change slightly.

The transient pressure wave will have a finite speed, which depends on the elasticity of the pipeline and fluid. This is illustrated by the example of field testing for water pipeline in Figure 6.

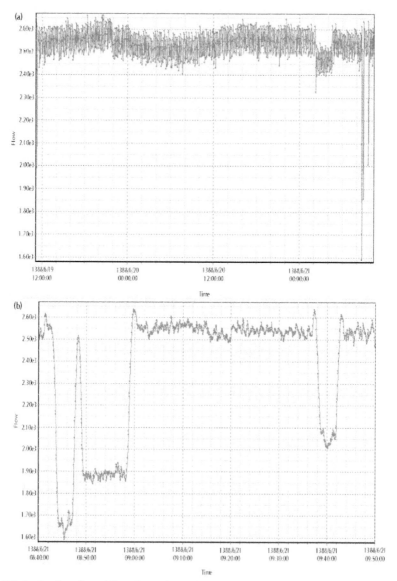

FIGURE 6   Registration of flow dynamics in time: (a) the entire length of pipe, (b) for the case of leakage during the water hammer.

Field tests were conducted on the actual water line. Using a regression model and the MOC for two cases described the separation of the water column and penetration of air into the pipeline.

(a) Pipeline in leakage condition, equipped with surge tank, air sucked in it at negative phase.

(b) Pipeline in leakage condition, not equipped with surge tank; there is no water column separation. This is consistent with the effect of the air outlet of the leak. The results showed that the pressure reaches with [6-7] (*atm.*). This pressure value is too high for the old pipeline, which should be viewed as a danger to the pipeline (for water with a surge tank and subject to leakage).

This case corresponds to the effect of withdrawal of water from the leak. Consequently, at 1 s was lost 494 l of water, which interred into the surge tank (water pipeline with a surge tank and subject to leakage). It was found that near the pumping station pressure wave reached 110 (*atm.*) at the beginning of water pipeline.

Water flow has decreased from 3,000 $\left(\frac{lit}{s}\right)$ to 2,500 $\left(\frac{lit}{s}\right)$.

This corresponds to the case of leakage in the water pipeline. The observation identifies transitional regime. Therefore, it detects the leakage by calculation. Thus, this work allows determining the location and amount of lost water. The modeling for calculations by the MOC was shown in Figure 7.

It is compared with the data of field tests (Table 1). In this work, for the calculation of water hammer software package have been used for three cases:

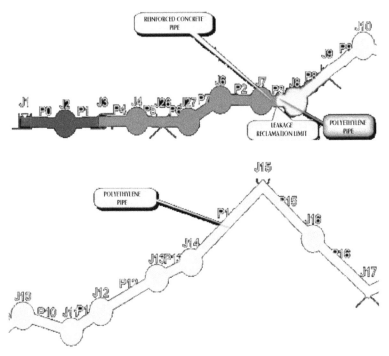

FIGURE 7   *(Continued)*

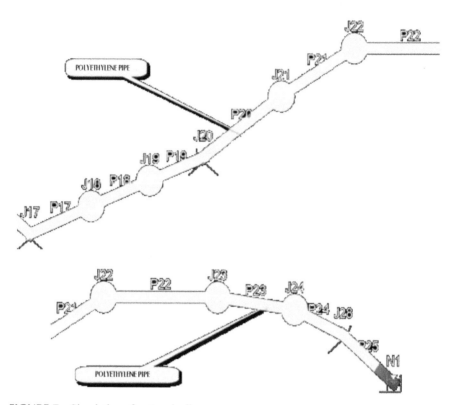

FIGURE 7   Simulation of water pipeline.

1. The first case, relates to the condition of water leakage in the water pipeline when the system is equipped with surge tank (real condition).
2. The second case, relates to the condition of water leakage in the water pipeline when the system is not equipped with surge tank.
3. The third case, relates to the condition of the absence of water leakage in the water pipeline in the absence of the surge tank [11].

TABLE 1   Model of field tests under extreme pressure and differential slope water.

End Point	Upsurge Ratio	Max. Pressure(m)	Min. Pressure (m)	Max. Head(m)	Min. Head(m)
P2:J7	1.25	120.5	96.7	156.8	133
P3:J7	1.25	120.5	96.7	156.8	133
P3:J8	1.25	118.7	95.1	154.8	131.2
P4:J3	1.1	109.4	99.5	144.9	135

TABLE 1   *(Continued)*

End Point	Upsurge Ratio	Max. Pressure(m)	Min. Pressure (m)	Max. Head(m)	Min. Head(m)
P4:J4	1.1	107.7	97.8	144.9	135
P5:J4	1.1	107.7	97.8	144.9	135
P5:J26	1.1	107.7	97.8	144.9	135
P6:J26	1.1	107.7	97.8	144.9	135
P6:J27	1.25	120.8	97	157.3	133.5
P7:J27	1.25	120.8	97	157.3	133.5
P7:J6	1.25	120.9	97	157.2	133.4
P8:J8	1.25	118.7	95.1	154.8	131.2
P8:J9	1.25	116.7	93.1	154.7	131.1
P9:J9	1.25	116.7	93.1	154.7	131.1
P9:J10	1.26	114.5	90.8	152.9	129.2
P10:J10	1.26	114.5	90.8	152.9	129.2
P10:J11	1.27	113.1	89.4	151.7	128
P11:J11	1.27	113.1	89.4	151.7	128
P11:J12	1.27	109.5	86.4	149.2	126
P12:J12	1.27	109.5	86.4	149.2	126
P12:J13	1.28	106.3	83.1	147.3	124.1
P13:J13	1.28	106.3	83.1	147.3	124.1
P13:J14	1.28	104.7	81.5	147	123.8
P14:J14	1.28	104.7	81.5	147	123.8
P14:J15	1.31	98.2	59.3	143.4	104.5
P15:J15	1.31	98.2	59.3	143.4	104.5
P15:J16	1.3	99.6	60.5	143	104
P16:J16	1.3	99.6	60.5	143	104
P16:J17	1.32	97.2	57.5	142.3	102.6
P17:J17	1.32	97.2	57.5	142.3	102.6
P17:J18	1.3	98.5	59.2	141.5	102.2
P18:J18	1.3	98.5	59.2	141.5	102.2

TABLE 1 *(Continued)*

End Point	Upsurge Ratio	Max. Pressure(m)	Min. Pressure (m)	Max. Head(m)	Min. Head(m)
P18:J19	1.3	99.1	60.3	141.4	102.7
P19:J19	1.3	99.1	60.3	141.4	102.7
P19:J20	1.31	97.2	57.9	141.4	102.1
P20:J20	1.31	97.2	57.9	141.4	102.1
P20:J21	1.3	95.1	56.5	137.6	99
P21:J21	1.3	95.1	56.5	137.6	99
P21:J22	1.32	91.8	52.9	136.4	97.5
P22:J22	1.32	91.8	52.9	136.4	97.5
P22:J23	1.52	66	24.5	135.9	97.5
P23:J23	1.52	66	24.5	135.9	97.5
P23:J24	1.71	53.9	12.2	135.7	97.5
P24:J24	1.71	53.9	12.2	135.7	97.5
P24:J28	2.07	36.4	5	131.6	97.5
P25:J28	2.07	36.4	5	131.6	97.5
P25:N1	1	16.7	16.7	112.6	97.5
P0:J1	0	0	0	40.6	40.6
P0:J2	1.09	5.5	4.7	41	40.2
P1:J2	1.09	5.5	4.7	41	40.2
P1:J3	1.12	5.7	4.5	41.2	40

For all three cases which were described, the output data were presented in the form of tables and graphs. These three cases have been proposed to study ways to modernize the surge tank. Comparison of the results for all three cases demonstrated the effectiveness of the surge tank during the entire transitional flow. Curves of transient flows include: flow-time, volume, time, pressure-time, pressure-time, minimum pressure-distance, increasing the distance, maximum-distance, and pressure-distance. In the final step, the final results (only for the first case–real condition) compared with the results of the first model under the assumption 2.

TABLE 2   Results of field trials in the water pipeline.

Pipe	Length(m)	Diameter(mm)	Velocity(m/s)	Hazen-Williams Friction Coef.
P3	311	1200	2.21	90
P4	1	1200	2.65	90
P5	0.5	1200	2.65	91
P6	108.7	1200	2.65	67
P7	21.5	1200	2.21	90
P8	15	1200	2.21	86
P9	340.7	1200	2.21	90
P10	207	1200	2.21	90
P11	339	1200	2.21	90
P12	328.6	1200	2.21	90
P13	47	1200	2.21	90
P14	590	1200	2.21	90
P15	49	1200	2.21	90
P16	224	1200	2.21	90
P17	18.4	1200	2.21	90
P18	14.6	1200	2.21	90
P19	12	1200	2.21	90
P20	499	1200	2.21	90
P21	243.4	1200	2.21	90
P22	156	1200	2.21	90
P23	22	1200	2.21	90
P24	82	1200	2.21	90
P25	35.6	1200	2.21	90
P0	0.5	1200	2.65	90
P1	0.5	1200	2.65	90

2.4   COMPARISON OF REGRESSION MODEL WITH THE METHOD OF CHARACTERISTICS MODEL_

Results obtained by regression analysis were compared in Figure 8 with the solution obtained by using the MOC model (second model-MOC).

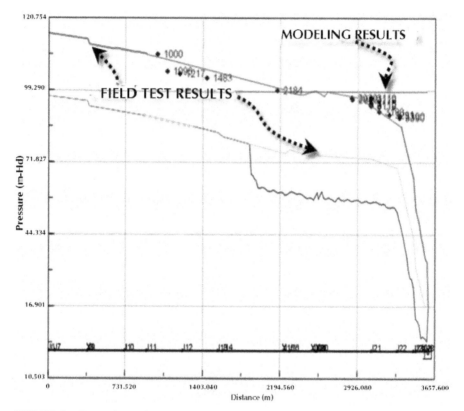

FIGURE 8 Comparison of simulation results (red line "method of characteristics", the dash line-method of regression analysis).

The Figure 8 shows that the results of calculations carried out by regression analysis almost coincide with the solution obtained by the MOC. Both methods calculations lead to the following conclusions (Table 3).

In leakage of water, the calculation results show that water loss was caused by a leak in a certain place of the water pipeline. Thus, there was a loss of water with a flow rate 494 $\left(\frac{lit}{s}\right)$, which interred in the tank (water pipeline with a surge tank, subject to leakage). It turned out that the surge wave formed near the pumping station (on start the transition curve).

Thus, water consumption decreased from 3,000 $\left(\frac{lit}{s}\right)$ to 2,500 $\left(\frac{lit}{s}\right)$. This served as an alarm signal for leakage. With the help of observations and calculations, tried for finding water leakage, location, rate and intensity [11].

2.4.1 Appearance of Negative Pressures

The profile of the minimum pressure curve lies under the profile of the transition curve, at the near of water reservoir. This indicates that in this zone line is faced by negative pressure. So it must be removal from the system. The maximum transient

TABLE 3 The simulation results of field tests in the water pipeline.

Node ID	Label	Category	Type	Elevation	X-cord.	Y-cord.	Branch pipes	Vapor pressure	Max. Volume	Type of volume	Code
1	J2	Junction	2 -more	35.5	7.32	0	2	-10	0	Vapor	68
2	J4	Junction	2 -more	37.2	20.52	0.34	2	-10	0	Vapor	68
3	J1	Reservoir	1 -more	40.6	0	0.3	1	-10	0	Vapor	23
4	J3	Pump	Shut	35.5	14.33	0.3	2	-10	0	Vapor	74
5	J7	Junction	2 -more	36.3	43.1	3.32	2	-10	0	Vapor	68
6	J8	Junction	2 -more	36.1	49.24	3.14	2	-10	0	Vapor	68
7	J11	Junction	2 -more	38.6	70.34	9.13	2	-10	0	Vapor	68
8	J12	Junction	2 -more	39.7	75.7	12.17	2	-10	0	Vapor	68
9	J14	Junction	2 -more	42.3	91.78	19.48	2	-10	0	Vapor	68
10	J16	Junction	2 -more	43.4	113.54	23.01	2	-10	0	Vapor	68
11	J18	Junction	2 -more	43	133.05	18.76	2	-10	0	Vapor	68
12	J19	Junction	2 -more	42.3	142.26	22.19	2	-10	0	Vapor	68
13	J21	Junction	2 -more	42.5	162.94	33.48	2	-10	0	Vapor	68
14	J22	Junction	2 -more	44.6	174.32	39.89	2	-10	0	Vapor	68
15	J23	Junction	2 -more	69.9	192.14	39.9	2	-10	0	Vapor	68
16	J27	Junction	2 -more	36.5	29.83	0.25	2	-10	0	Vapor	68

TABLE 3 *(Continued)*

Node ID	Label	Category	Type	Elevation	X-cord.	Y-cord.	Branch pipes	Vapor pressure	Max. Volume	Type of volume	Code
17	J26	Prot equip	air valve	37.2	25.22	0.16	2	–10	0	Air	76
18	J9	Prot equip	air valve	38	55.2	7.47	2	–10	0	Air	76
19	J10	Junction	2 -more	38.3	61.71	11.54	2	–10	0	Vapor	68
20	J15	Prot equip	air valve	45.2	104.6	30.77	2	–10	0	Air	76
21	J17	Prot equip	air valve	45	123.39	15.06	2	–10	0	Air	76
22	J20	Prot equip	air valve	44.2	150.66	25.26	2	–10	0	Air	76
23	J24	Junction	2 -more	81.8	204.22	38.36	2	–10	0	Vapor	68
24	J28	Prot equip	air valve	95.2	211.71	35.02	2	–10	198.483	Air	76
25	N1	Reservoir	1 -more	95.9	219.75	28.79	1	–10	0	Vapor	23
26	J6	Junction	2 -more	36.3	35.7	3.5	2	–10	0	Vapor	68
27	J13	Junction	2 -more	41	85.73	17.14	2	–10	0	Vapor	68
24	J28	Protection equip	air valve	95.2	211.71	35.02	2	–10	198.483	Air	76

pressure line lies entirely above the equilibrium line of pressure flow curve. Maximum system pressure reached 156.181 *(m)*.

It is too high pressure for an old pipeline, and should be assumed as a hazard to the pipeline system. Pressure-time curve for the water line without a surge tank under the condition of leakage shows that for 1.2 s the pressure value increased from 1.2 *(m)* to 146 *(m)*.

Also it decreased at 1.5 s to 131 *(m)*. Then within 5 s, pressure value has increased to 135 *(m)*.

Air Entrance Approaches

Analysis and comparison of the first and the second model showed that at point P24: J28, air penetrated into water pipeline .The maximum amount of infiltrated air was equal to 198,483 (m^3), and the flux was equal to 2,666 $\left(m^3 \big/ s \right)$.

Comparing the results of a regression model and the MOC showed the influence of the surge tank on total transmitted flow in pipeline (depending on the flow-time, the pressure-time and pressure-distance curves).

It showed the effective role of the surge tank. The flow was reduced from 3,014 $\left(lit \big/ s \right)$ to the minimum value of 2,520 $\left(lit \big/ s \right)$ for 6 s. So for 4 s, it increased to 3,228 $\left(lit \big/ s \right)$ It is the effect of local diversion of the general flow of the pipeline. It is investigated the effect of the oscillation period by the ratio of discharge from the local diversion to the full transmitted flow at the pipeline. Subsequently, it revealed the effect of the amplitude of the wave, if there was an outflow of the pool at high-pressure leak.

Effect of local diversion on a maximum pressure is the reason for the high pressure in the water supply pipeline. It is connect with the effect of leakage (the effect of local leakage at high pressure, decaying in the water).

This can explain the reasons of repeated ruptures of the pipe. Leakage occurs near the water treatment plant. Thus, the water flow was reduced from 3,000 $\left(lit \big/ s \right)$ to 2,500 $\left(lit \big/ s \right)$.

TABLE 4 Model of extreme pressures for pipes (provided by water hammer).

Point	Distance	Elev.	Init Head	Max Head	Min Head	Vapor Pr.	
+ P3:J7	.0	36.3	132.8	156.8	105.1	.000	−10.0
P3:5.00%	15.6	36.3	132.7	156.6	105.3	.000	−10.0
P3:10.00%	31.1	36.3	132.6	156.3	105.1	.000	−10.0
P3:15.00%	46.7	36.3	132.4	156.1	105.0	.000	−10.0
P3:20.00%	62.2	36.3	132.3	155.8	104.9	.000	−10.0
P3:25.00%	77.8	36.3	132.2	155.6	104.7	.000	−10.0
P3:30.00%	93.3	36.2	132.1	155.5	104.4	.000	−10.0
P3:35.00%	108.9	36.2	131.9	155.3	104.3	.000	−10.0
P3:40.00%	124.4	36.2	131.8	155.2	104.2	.000	−10.0

TABLE 4 *(Continued)*

Point	Distance	Elev.	Init Head	Max Head	Min Head	Vapor Pr.	
P3:45.00%	140.0	36.2	131.7	155.0	104.0	.000	−10.0
P3:50.00%	155.5	36.2	131.6	135.1	103.8	.000	−10.0
P3:55.00%	171.1	36.2	131.4	134.9	103.7	.000	−10.0
P3:60.00%	186.6	36.2	131.3	134.7	103.5	.000	−10.0
P3:65.00%	202.2	36.2	131.2	134.4	103.4	.000	−10.0
P3:70.00%	217.7	36.2	131.1	134.6	102.9	.000	−10.0
P3:75.00%	233.3	36.2	130.9	134.4	102.7	.000	−10.0
P3:80.00%	248.8	36.1	130.8	134.0	102.7	.000	−10.0
P3:85.00%	264.4	36.1	130.7	133.9	102.4	.000	−10.0
P3:90.00%	279.9	36.1	130.6	133.6	102.3	.000	−10.0
P3:95.00%	295.5	36.1	130.4	133.5	102.1	.000	−10.0

The pipeline was equipped with a 50,000 (m^3) reservoir. The leak occurred at the point P3: J7 (% P3: 45.00) at an elevation of 36.2 (m) and a distance of 140 (m) far away from the water pumping station.

At the point P3: J7, maximum head of 155 (m) fell to 135.1 (m). The vapor pressure was equal to –10.0 (m). The initial and minimum pressure did not change (Table 3 and Table 4).

This was the alarm of water hammer. Therefore, the advanced modeling techniques and numerical analysis for founding the location and rate of the water losses from the water pipeline. In the final stage, similar studies were conducted for all water systems [12].

TABLE 5 High-speed sensors (field trials) for hydraulic shock.

Pressure	flow	Distance	time	Pressure
(m-Hd)	(lit/s)	(m)	(s)	(m-Hd)
Field				Model
86	2491	3390	0	89.033
86	2491	3390	1	89.764
88	2520	3291	0	90.427
90	2520	3190	1	91.663
95	2574	3110	1.4	94.03
95	2574	3110	1.4	94.03
95	2574	3110	1.5	94.1
95	2590	3110	2	94.96

TABLE 5 *(Continued)*

Pressure	flow	Distance	time	Pressure
(m-Hd)	(lit/s)	(m)	(s)	(m-Hd)
Field				Model
95	2590	3110	2	94.96
95.7	2600	3110	2	95.27
95.7	2600	3110	3	96.73
95.7	2600	3110	4	96.73
95.7	2600	3110	5	97.46
95.7	2605	3110	0.5	94.33
100	2633	2184	1.3	100.41
100	2633	2928	1.3	96.69
101	2650	2920	1.4	97.33
106	2680	1483	1.4	105.45
107	2690	1217	1.4	107.09
109	2710	1096	1.4	108.31
109	2710	1096	1.4	108.31
110	2920	1000	1.5	115.37

Simulation and methodology were compared with the results of other authors' studies. Generally inaccurate representations of friction force, steam cavitation and boundary conditions will be lead to errors. Gray, 1953, Streeter and Lai [13], Elansary, Silva, and Chaudhry [14-15] worked in these fields. Data were collected for the water pipeline (pilot field testing) based on GIS and regression analysis with a fairly low correlation coefficient around the scattering pattern. Observation and records were made by computerized system named "PLC".

At water pipeline high speed sensors were applied for accurate measurement of data values or to track transients (Table 5).

The simulation were performed by several methods. These methods provided an indication of the reliability of the results. The characteristic and main advantage was focus on a new idea. It has found the location of losses in water supply pipelines.

2.4.2 Means of Protection Against Water Hammer

Protection of hydraulic systems by releasing part of the transported fluid is the most widespread method of artificial reduction of the hydraulic shock. Devices that perform this function can be divided into valve, bursting disc and the overflow of the column. Overflow column due to high pressure and large geodetic height of the systems as not applicable. Valve device for protection against water hammer can be divided into safety valves, and special shock absorbers are hydraulic shocks. Safety valves of all types have a number of specific deficiencies.

This is a big pressure difference between the opening and closing valves, different slamming the gate and generate an additional blow to the moment approach of negative pressure wave and the consequent need for test positives. Persistence settings of safety valves prevents quenching pressure shocks, starting with lowering the pressure, because they do not react to the effect of lowering the pressure in the pipelines at the time of the negative pressure wave.

All this which was resulted in dewatering facilities are not adopted in piping hydraulic systems. Recently, more and more widely used designs, responsive to the positive derivative of pressure over time, due, primarily, their universality. Most of these devices, direct action and, therefore, the stop-body is both a sensitive element. This contributes to high speed devices.

It able to easily provide the minimum necessary to absorb the impact of transported fluid drain. At the same time, there is dependence between sensitivity and sealing force of the valve. It leads to the fact that the required sensitivity of the device determines the design effort in a valve pair, which is insufficient. Free of these shortcomings there are devices for indirect actions.

They will contain a link gain that is included the measuring elements and the executive body. All known constructions of hydraulic shock absorbers indirect action (often referred to as impulsive) contain the sensing element.

It is used as a spring-loaded safety valve, lever-piston system, and solenoid valve. It acts on the control element, which often looks like a valve or a pair of hostages. Managing element informs the cavity of the hydraulic drive.

It relates to the atmosphere or with the pressure line and thereby opens or closes shut off damper.

The area of the piston of the hydraulic drive can be considerably greater than the area drain valve. In general, hydraulic shock absorbers have the advantages of indirect action. As an opportunity it obtains the necessary sealing force to achieve a reliable seal shut-off body.

It has relatively small size, even for a very powerful hydraulic system. It achieved the required sensitivity can identify them as the most promising among the devices. These devices protect the hydraulic system by resetting of the transported fluid. In this system, pumping applies hydraulic impact damper with air-hydraulic valve. Application of air-hydraulic valves as a means to combat water hammer is widely used in hydraulic engineering.

It used especially in systems of hydraulic transport, where the valve device may not operate reliably due to the presence in the flow of solids. The valves are easy to design and necessary (in accordance volume of compressed air hydraulic parameters) and protection against pressure fluctuations, including high.

In practice, there are possible the use of other methods for protection against water hammer. There are air-inlet and a fluid drain through the pump, check valves of the pipeline section, the application of equalization tanks, interference suppressors, the introduction into the pipeline of elastic elements, and so on.

The method of calculation and means of protection against water hammer was developed. It improves standards in the design and construction works in excess of atmospheric pressure at water pipelines. Implementation of research findings contributed to maintain reliable transmission of water through the main pipeline (Figure 9).

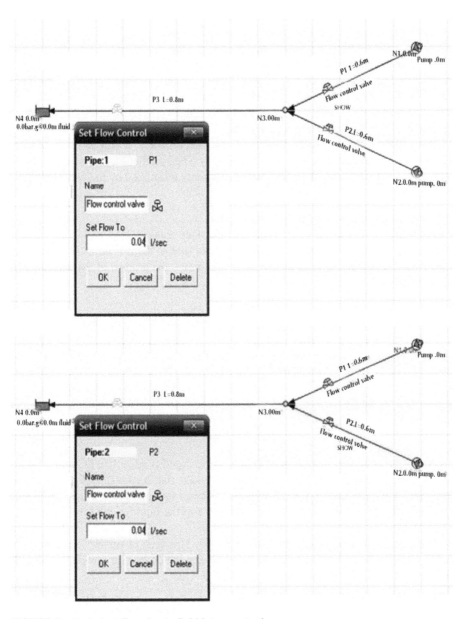

FIGURE 9   Turbulent flow due to fluid interpenetration.

2.5 PERTURBATIONS IN HOMOGENEOUS AND STRATIFIED FLOWS

One of the problems in the study of fluid flow in plumbing systems is the behavior of stratified fluid in the channels. Mostly steady flows initially are ideal, then the viscous and turbulent fluid in the pipes.

If you look at the deep pool filled with water, and on its surface to create a disturbance, then the surface of the water will begin to propagate. Their origin is explained by the fact that the fluid particles are located near the cavity.

They create disturbances which will seek to fill the cavity under the influence of gravity. The development of this phenomenon is led to the spread of waves on the water. The fluid particles in such a wave do not move up and down around in circles. The waves of water are neither longitudinal nor transverse. They seem to be a mixture of both. The radius of the circles varies with depth of moving fluid particles. They reduce to as long as they do not become equal to zero.

If we analyze the propagation velocity of waves on water, it will be revealed that the velocity of waves depends on length of waves. The speed of long waves is proportional to the square root of the acceleration of gravity multiplied by the wave length:

$$v_\phi = \sqrt{g\lambda}$$

The cause of these waves is the force of gravity.

For short waves the restoring force due to surface tension force, and therefore the speed of these waves is proportional to the square root of the private. The numerator of which is the surface tension, and in the denominator, the product of the wavelength to the density of water:

$$v_\phi = \sqrt{\sigma / \lambda\rho}$$

Suppose there is a channel with a constant slope bottom, extending to infinity along the axis Ox. And let the feed in a field of gravity flows, incompressible fluid. It is assumed that the fluid is devoid of internal friction. Friction neglects on the sides and bottom of the channel. The liquid level is bottom of the channel h. A small quantity compared with the characteristic dimensions of the flow, the size of the bottom roughness, and so on.

Let $h = \xi + h_0$

where h^0 = ordinate denotes the free surface of the liquid (Figure 10). Free liquid surface h_0, which is in equilibrium in the gravity field is flat. As a result of any external influence, liquid surface in a location removed from its equilibrium position. There is a movement spreading across the entire surface of the liquid in the form of waves, called gravity.

They are caused by the action of gravity field. This type of waves occurs mainly on the liquid surface. They capture the inner layers, the deeper for the smaller liquid surface.

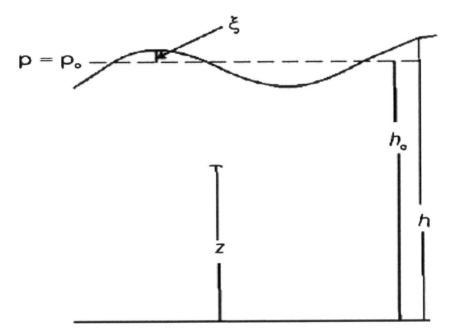

FIGURE 10 Schematic showing the layer of fluid of variable depth.

where h_0 is the level of the free surface,

$\xi = $ A deviation from the level of the liquid free surfac

$h = $ Depth of the fluid, and

$z = $ Vertical coordination of any point in the water column. We assume that the fluid flow is characterized by a spatial variable x and time dependent t.

Thus, it is believed that the fluid velocity u has a nonzero component u_x which will be denoted by u (other components can be neglected in addition, the level of h depends only on x and t.

Let us consider such gravitational waves, in which the speed of moving particles are so small that for the Euler equation, it can be neglected the $(u\nabla)u$ compared with $\partial u / \partial t$.

During the time period τ, committed by the fluid particles in the wave, these particles pass the distance of the order of the amplitude a.

Therefore, the speed of their movement will be $u \sim a/\tau$.

Rate u varies considerably over time intervals of the order τ and for distances of the order λ along the direction of wave propagation,

$\lambda = $ Wavelength.

Therefore, the derivative of the velocity time order u/τ and the coordinates order u/λ

Thus, the condition:

$$(u\nabla)u << \partial u / \partial t$$

Equivalent to the requirement

$$\frac{1}{\lambda}\left(\frac{a}{\lambda}\right)^2 << \frac{a}{\tau}\frac{1}{\tau} \qquad\qquad a << \lambda, \qquad \text{or,} \qquad (2.4.1)$$

That is amplitude of the wave must be small compared with the wavelength. Consider the propagation of waves in the channel Ox directed along the axis for fluid flow along the channel.

Channel cross section can be of any shape and change along its length with changes in liquid level, cross-sectional area of the liquid in the channel denoted by: $h = h(x,t)$.

The depth of the channel and basin are assumed to be small compared with the wavelength.

We write the Euler equation in the form of:

$$\frac{\partial u}{\partial t} = -\frac{1}{\rho}\frac{\partial p}{\partial x} \qquad\qquad (2.4.2)$$

$$\frac{1}{\rho}\frac{\partial p}{\partial z} = -g \qquad\qquad (2.4.3) \text{ where } \rho = \text{Density,}$$

p = Pressure,
g = Acceleration of free fall.

Quadratic in velocity members omitted, since the amplitude of the waves is still considered low.

From the second equation we have that at the free surface:

$$z = h(x,t)$$

where, $p = p_0$ should be satisfied:
$p = p_0 + \rho g(h - z)$ 2.4.4) Substituting this expression in Equation (2.4.2), we obtain:

$$\frac{\partial u}{\partial t} = -g\frac{\partial h}{\partial x} \qquad\qquad (2.4.5)$$

To determine u and h we use the continuity equation for the case under consideration.

Consider the volume of fluid contained between two planes of the cross-section of the canal at a distance dx from each other per unit time through a cross-section x enter the amount of fluid, equal to $(hu)_x$.

At the same time through the section:

$$x + dx$$

There is forth coming $(hu)_{x+dx}$.

Therefore, the volume of fluid between the planes is changed to:

$$(hu)_{x+dx} - (hu)_x = \frac{\partial(hu)}{\partial x}dx$$

By virtue of incompressibility of the liquid is a change could occur only due to changes in its level. Changing the volume of fluid between these planes in a unit time is equal $\frac{\partial h}{\partial t} dx$ Consequently, we can write:

$$\frac{\partial(hu)}{\partial x}dx = -\frac{\partial h}{\partial t}dx \quad \text{and} \quad \frac{\partial(hu)}{\partial x}+\frac{\partial h}{\partial t}=0, \; t>0, \; -\infty < x < \infty \text{ or,} \quad (2.4.6)$$

Since, $h = h_0 + \xi$ where a h_0 = denotes the ordinate of the free liquid surface, in a state of relative equilibrium and evolving the influence of gravity is

$$\frac{\partial \xi}{\partial t}+h_0\frac{\partial u}{\partial x}=0 \quad (2.4.7)$$

Thus, we obtain the following system of equations describing the fluid flow in the channel:

$$\frac{\partial \xi}{\partial t}+h_0\frac{\partial u}{\partial x}=0, \; \frac{\partial u}{\partial t}+g\frac{\partial \xi}{\partial x}=0, \; t>0, \; -\infty < x < \infty \quad (2.4.8)$$

2.5.1 Velocity Phase of the Harmonic Wave

The phase velocity h_0 expressed in terms of frequency v_Φ and wavelength f (or the angular frequency) λ and wave number $\omega = 2\pi f$ formula $k = 2\pi/\lambda$.

The concept of phase velocity can be used if the harmonic wave propagates without changing shape.

This condition is always performed in linear environments. When the phase velocity depends on the frequency, it is equivalent to talk about the velocity dispersion. In the absence of any dispersion the waves assumed with a rate equal to the phase velocity.

Experimentally, the phase velocity at a given frequency can be obtained by determining the wavelength of the interference experiments. The ratio of phase velocities in the two media can be found on the refraction of a plane wave at the plane boundary of these environments. This is because the refractive index is the ratio of phase velocities.

It is known that the wave number k satisfies the wave equation are not any values ω but only if their relationship. To establish this connection is sufficient to substitute the solution of the form:

$\exp[i(\omega t - kx)]$ in the wave equation.

The complex form is the most convenient and compact. We can show that any other representation of harmonic solutions, including in the form of a standing wave leads to the same connection between ω and k .

Substituting the wave solution into the equation for a string, we can see that the equation becomes an identity for:

$$\omega^2 = k^2 v_\phi^2$$

Exactly the same relation follows from the equations for waves in the gas, the equations for elastic waves in solids and the equation for electromagnetic waves in vacuum. The presence of energy dissipation [16] leads to the appearance of the first derivatives (forces of friction) in the wave equation. The relationship between frequency and wave number becomes the domain of complex numbers. For example, the telegraph equation (for electric waves in a conductive line) yields:

$$\omega^2 = k^2 v_\phi^2 + i \cdot \omega R / L$$

The relation connecting between a frequency and wave number (wave vector), in which the wave equation has a wave solution is called a dispersion relation, the dispersion equation or dispersion. This type of dispersion relation determines the nature of the wave. Since the wave equations are partial differential equations of second order in time and coordinates, the dispersion is usually a quadratic equation in the frequency or wave number.

The simplest dispersion equations presented above for the canonical wave equation are also two very simple solutions:

$$\omega = +k v_\phi \text{ and } \omega = -k v_\phi$$

We know that these two solutions represent two waves traveling in opposite directions. By its physical meaning the frequency is a positive value so that the two solutions must define two values of the wave number, which differ in sign. The act permits the dispersion, generally speaking, the existence of waves with all wave numbers that is of any length, and, consequently, any frequencies. The phase velocity of these waves:

$$v_\Phi = \omega / k$$

Coincides with the most velocity, which appears in the wave equation and is a constant which depends only on the properties of the medium.

The phase velocity depends on the wave number, and, consequently, on the frequency. The dispersion equation for the telegraph equation is an algebraic quadratic equation has complex roots. By analogy with the theory of oscillations, the presence of imaginary part of the frequency means the damping or growth of waves. It can be noted that the form of the dispersion law determines the presence of damping or growth.

In general terms, the dispersion can be represented by the equation:

$$\Phi(\omega, k) = 0$$

where Φ = A function of frequency and wave vector.

By solving this equation for ω you can obtain an expression for the phase velocity):

$$v_\Phi = \omega / k = f\left(\omega, \vec{k}\right).$$

By definition, the phase velocity is a vector directed normal to phase surface.

Then, more correctly write the last expression in the following form:

$$\vec{v}_\Phi = \frac{\lambda}{T} = \frac{\omega}{k^2} \cdot \vec{k} = f\left(\omega, \vec{k}\right)$$

2.5.2   Dispersive Properties of Media

The most important subject of research in wave physics, which has the primary practical significance.

If we refer to dimensionless parameters and variables.

$$\tau = t\sqrt{\frac{g}{h_0}}, \quad X = \frac{x}{h_0}, \quad U = u\frac{1}{\sqrt{gh_0}}, \quad \delta = \frac{\xi}{h_0},$$

The system of Equations (2.4.8) becomes

$$\frac{\partial \delta}{\partial \tau} + \frac{\partial U}{\partial X} = 0, \quad \frac{\partial U}{\partial \tau} + \frac{\partial \delta}{\partial X} = 0, \quad t > 0, \quad -\infty < X < \infty \tag{2.4.9}$$

Consider a plane harmonic longitudinal waves, that is., we seek the solution of (2.4.9) as the real part of the following complex expressions:

$$\Psi = \Psi^0 \exp[i(k_* X + \omega_* \tau)], \quad \Psi^0 = \Psi_*^0 + i\Psi_{**}^0, \quad |\Psi^0| << 1$$

$$k_* = k + ik_{**}, \quad \omega_* = \omega + i\omega_{**} \tag{2.4.10}$$

where, determines the amplitude of the perturbations of displacement and velocity. $\Psi = \delta, U$, a $\Psi^0 = \delta^0, U^0$

There are two types of solutions:

Type I: Solution or wave of the first type, when:
$k_* = k = $ A real positive number $(k > 0, k_{**} = 0)$.
In this case, we have:

$$\Psi = \left(\Psi_*^0 + i\Psi_{**}^0\right)\exp[i(kX + \omega\tau + i\omega_{**}\tau)] = \left(\Psi_*^0 + i\Psi_{**}^0\right)\exp(-\omega_{**}\tau)\times$$

$$[\cos(kX + \omega\tau) + i\sin(kX + \omega\tau)]$$

$$\text{Re}\{\Psi\} = \exp(-\omega_{**}\tau)\Psi^0 \sin[\varphi + (kX + \omega\tau)]$$

$$|\Psi^0| = \sqrt{\Psi_*^{0^2} + \Psi_{**}^{0^2}}, \quad \varphi = arctg\left(-\Psi_*^0 / \Psi_{**}^0\right)$$

Thus, the decision of the first type is a sinusoidal coordinate and $\omega_{**} > 0$ decaying exponentially in time perturbation, which is called k = wave:

$$\Psi(k) = |\Psi^0| \exp[-\omega_{**}(k)\tau]\sin\left\{\varphi + \frac{2\pi[X + v_\phi(k)\tau]}{\lambda(k)}\right\} \tag{2.4.11}$$

where,
$(v_\phi(k) = \omega(k)/k, \lambda(k) = 2\pi/k), \varphi), \varphi = $ Initial phase.
Here,
$v_\phi(k) = $ phase velocity or the velocity of phase fluctuations,
$\lambda(k) = $ Wavelength,

$\omega_{**}(k)$ = damping the oscillations in time.

In other words, k = Waves have uniform length, but time-varying amplitude.

These waves are analogue of free oscillations.

Type II: Decisions, or wave, the second type, when:

$$\omega_* = \omega = a$$

Real positive number $(\omega > 0, \omega_{**} = 0)$.

In this case, we have;

$$\psi = (\psi_*^0 + i\psi_{**}^0)\exp[i(kX + \omega\tau + ik_{**}z)] = (\Psi_*^0 + i\Psi_{**}^0)\exp(-k_{**}X)\times$$
$$[\cos(kX + \omega\tau) + l\sin(kX + \omega\tau)]$$

Thus, the solution of the second type is a sinusoidal oscillation in time (excited, for example, any stationary source of external monochromatic vibrations at) $X = 0$, decaying exponentially along the length of the amplitude.

Such disturbances, which are analogous to a wave of forced oscillations, called ω is waves:

$$\Psi(\omega) = |\Psi^0(\omega)|\exp(-k_{**}(\omega)X)\sin\left\{\varphi + \frac{2\pi[X + v_\phi(\omega)\tau]}{\lambda(\omega)}\right\}$$

$$v_\phi(\omega) = \omega/k(\omega),$$

$$\lambda(\omega) = 2\pi/k(\omega)$$

(2.4.12)

Here, $k_{**}(\omega)$ = damping vibrations in length.

In other words, ω = waves–waves with stationary in time but varying in length amplitudes.

Cases $k < 0, k_{**} > 0$ and $k > 0, k_{**} < 0$ consistent with attenuation of amplitude the disturbance regime in the direction of phase fluctuations or phase velocity.

Let us obtain the characteristic equation, linking k_* and ω_*.

After substituting (2.4.10) in the system of Equation (2.4.9) we obtain:

$$\delta^0\frac{\omega_*}{k_*} + U^0 = 0, \qquad U^0\frac{\omega_*}{k_*} + \delta^0 = 0$$

(2.4.13)

From the condition of the existence of a system of linear homogeneous algebraic Equation (2.4.13) with respect to perturbations of a nontrivial solution implies the desired characteristic, or dispersion, which has one solution:

$$v_\phi = \sqrt{gh_0}$$

(2.4.14)

Thus, we obtain a solution representing a sinusoidal in time and coordinate free undammed oscillations. Such behaviors of the waves are due to the absence of any dissipation in the fluid. The fluid is incompressible and ideal. There is no heat mass transfer.

Equation (2.4.9) with respect to perturbations take the form of wave equations:

$$\frac{\partial^2 \xi}{\partial t^2} = gh_0 \frac{\partial^2 \xi}{\partial x^2} \qquad \text{and} \qquad \frac{\partial^2 u}{\partial t^2} = gh_0 \frac{\partial^2 u}{\partial x^2} \qquad (2.4.15)$$

Note that in gas dynamics $v_\phi = \sqrt{gh_0}$ equivalent to the speed of sound.

KEYWORDS

- **Dewatering systems**
- **Geography information systems**
- **Hydralic system**
- **Pipe wall**
- **Program logic control**

REFERENCES

1. Kraichnan, R. H. and Montgomery, D. Two-dimensional turbulence, *Rep. Prog. Phys.*, *43*, *547* 1417–1423 (1967).
2. Tullis, J. P. *Control of Flow in Closed Conduits.* Fort Collins, Colorado, pp. 315–340 (1971).
3. Wood, D. J., Dorsch, R. G., and Lightener, C. Wave-Plan Analysis of Unsteady Flow in Closed Conduits. *Journal of Hydraulics Division, ASCE, 92*, 83–110 (1966).
4. Wood, D. J., and Jones, S. E. *Water hammer charts for various types of valves.* Journal of the Hydraulics Division, Proceedings of the American Society of Civil Engineers, pp. 167–178 (January, 1973).
5. Leon Arturo, S. An efficient second-order accurate shock-capturing scheme for modeling one and two-phase water hammer flows, PhD Thesis, pp. 4–44 (March 29, 2007)
6. Vallentine, H. R. Rigid Water Column Theory for Uniform Gate Closure, *J. of Hyd. Div. ASCE*, pp. 55–243 (July, 1965).
7. Watters, G. Z., *Modern Analysis and Control of Unsteady Flow in Pipelines.* Ann Arbor Sci., 2nd Ed., pp. 1098–1104 (1984).
8. Tennekes, H., and Lumley, J. L. *A First Course in Turbulence.* MIT Press, pp. 410–66 (1972).
9. Thorley, A. R. D. *Fluid Transients in Pipeline Systems.* D&L. George Herts, England, pp. 231–242 (1991).
10. Lee, T. S, and Pejovic, S. Air influence on similarity of hydraulic transients and vibrations. *ASME J. Fluid Eng., 118*(4), 706–709 (1996).
11. Hariri Asli, K., Nagiyev, F. B., and Haghi, A. K. *Water hammer and fluid condition.* 9th Conference on Ministry of Energetic works at research week, Tehran, Iran, pp. 27–43, http://isrc.nww.co.ir. (2008).
12. Choi, S. B., Barren, R. R., and Warrington R. O. *Fluid Flow and Heat Transfer in Micro-tubes.* ASME DSC *40*, pp. 89–93 (1991).
13. Moody, L. F. Friction Factors for Pipe Flow. *Trans. ASME, 66*, 671–684 (1944).
14. Bergeron, L. *Water hammer in Hydraulics and Wave Surge in Electricity.* John Wiley & Sons, Inc., New York, pp. 102–109 (1961).

15. Bracco, A., McWilliams, J. C., Murante, G., Provenzale, A., and Weiss, J. B. Revisiting freely decaying two-dimensional turbulence at millennial resolution. *Phys. Fluids, Issue, 12*(11) 2931–2941 (2000).
16. Loytsyanskiy, L. G. *Fluid.* Nauka, Moscow, p. 904 (1970).

3 Heat Transfer and Phase Transitions in Binary Mixtures of Liquids with Vapor Bubble

CONTENTS

3.1   TYPES AND COMBINATIONS OF TWO PHASE FLOW

In the two phase flow is extremely important to the concept of volume concentration. This is the relative volume fraction of one phase in the volume of the pipe [1]. Such an environment typical fluid is a high density and little compressibility. This property contributes to the creation of various forms of transient conditions [2-3]. Figure 1 shows an experimental setup, which investigated the formation of different modes of fluid flow with gas bubbles and steam. The experimental results show that the bubble flow usually occurs at low concentrations of vapor. It includes three main types of flow regimes in microgravity bubble, slug and an annular (Figure 2).

FIGURE 1 Snapshot laboratory setup for studying the structure of flow in different configurations tube.

Bubble Flow

FIGURE 2 *(Continued)*

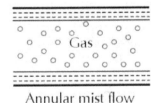

Slug Flow

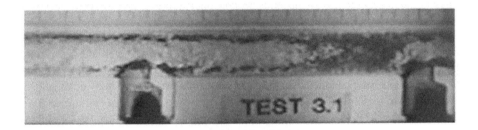

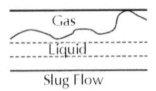

Annular mist flow

Stratified Flow

Wavy Stratified Flow

FIGURE 2   Types of flows.

Figure 2 shows snapshots of the flow pattern for various configurations of the tube [4-5]. The flow enters the tee at the bottom of the picture, and then is divided at the entrance to the tube. The inner diameter of the tee is 1.27 cm (Figure 1.).The narrowing of the flow is achieved by reducing the diameter of the hose.

Within the reduction is a liquid recirculation zone, called the "vena contraction" Wet picture that, when the liquid is re-circulated to the "vena contraction". However, there are conditions, whereby the gas phase of the contract is caught vena. Figure 3 shows the different flow regimes observed in the experimental setup [6].

Increased flow is achieved by increasing the size of the pipe. Again, there is an area of the liquid recirculation near the "corner" a sudden expansion. Depending on the level of consumption of bubbly liquid or gas, it falls into the trap in this area.

In the inlet fluid moves out of the pipe diameter of 12.7 (*mm*) in the pipe diameter of 25 (*mm*).

Normal extension occurs at the beginning of the flow. Soon comes the expansion section, and the flow rate continues to increase. Two phase jet stream created, ultimately, with areas of air flowing above and below the bubble region [6]. The behavior of gas-liquid mixture in the expansion is proportional to increasing the diameter of the pipe [7-8].

It is shown that in place of the sudden expansion of a transition flow regime of the turbulent flow. Depending on the flow or gas bubble mixture it falls into the trap in this area.

Experiments were conducted on the pipe, whose diameter is suddenly doubled [9]. In this case the region of turbulence of fluid are observed around the "corner" a sudden expansion. The expansion is observed at the beginning of the flow. As a result of turbulence flow gap expansion increases and the flow rate continues to increase. In the end, creates a stream of two phase flow, with air fields, the current above and below the bubble area [10-11].

With this experimental setup is shown that the formation of different modes of two phase flow depends on the relative concentration of these phases and the flow rate.

Figure 1 and Figure 2 shows a diagram of core flow of vapor-liquid flow regimes, in particular, the bubble, stratified, slug, stratified, and the wave dispersion circular flow.

In these experiments the mode of vapor-liquid flow in a pipe, when the bubbles are connected in long steam field, whose dimensions are commensurate with the diameter of the pipe [12-13].

This flow is called the flow of air from the tube. In the transition from moderate to high-speed flow, when the concentration of vapor and liquid are approximately equal, the flow regime is often irregular and even chaotic [14-15]. By the simulated conditions, It is assumed that the electricity suddenly power off without warning (i.e., no time to turn the diesel generators or pumps) [15-16].

Such situations are the strong reason of the installation of pressure sensors, equipped with high speed data loggers. Therefore, the following items are consequences which may result in these situations.

Normal

Gap

Jet

FIGURE 3   Narrowing and sudden expansion flow level.

3.1.1 Effects of Transients

Hydraulic transients can lead to the following physical phenomenaHigh or low transient pressures that may arise in the piping and connections in the share of second. They often alternate from highest to lowest levels and *vice versa*.

High pressures are a consequence of the collapse of steam bubbles or cavities are similar to steam pump cavitations. It can yield the tensile strength of the pipes. It can also penetrate the groundwater into the pipeline [17-18].

3.1.2 High Transient Flows

High-speed flows are also very fast pulse pressure. It leads to temporary but very significant transient forces in the bends and other devices that can make a connection to deform. Even strain buried pipes under the influence of cyclical pressures may lead to deterioration of joints and lead to leakage. In the low-pressure pumping stations at downstream a very rapid closing of the valve, known as shut off valve, may lead to high pressure transient flows.

3.1.3 Water Column Separation

Water column, usually are separated with sharp changes in the profile or the local high points. It is because of the excess of atmospheric pressure. The spaces between the columns are filled with water or the formation of steam (e.g., steam at ambient temperature) or air, if allowed admission into the pipe through the valve. Collapse of cavitation bubbles or steam can cause the dramatic impact of rising pressure on the transition process.

If the water column is divided very quickly, it could in turn lead to rupture of the pipeline. Vapors cavitation may also lead to curvature of the pipe. High pressure wave can also be caused by the rapid removal of air from the pipeline.

Steam bubbles or cavities are generated during the hydraulic transition. The level of hydraulic pressure (EGD) or pressure in some areas could fall low enough to reach the top of the pipe. It leads to sub-atmospheric pressure or even full-vacuum pressures. Part of the water may undergo a phase transition, changing from liquid to steam, while maintaining the vacuum pressure.

This leads to a temporary separation of the water column. When the system pressure increases, the columns of water rapidly approach to each other. The pair reverts to the liquid until vapor cavity completely dissolved. This is the most powerful and destructive power of water hammer phenomenon.

3.1.4 Global Regulation of Steam Pressure

If system pressure drops to vapor pressure of the liquid, the fluid passes into the vapor, leading to the separation of liquid columns. Consequently, the vapor pressure is a fundamental parameter for hydraulic transient modeling. The vapor pressure varies considerably at high temperature or altitude.

Fortunately, for typical water pipelines and networks, the pressure does not reach such values. If the system is at high altitude or if it is the industrial system, operating at high temperatures or pressures, it should be guided by a table or a state of vapor pressure curve vaporliquid.

3.1.5   Vibration

Pressure fluctuations associated with the peculiarities of the system, as well as the peculiarities of its design. The pump must be assumed just as one of the promoters of the system. Effect of pressure relief valve on the fluctuations of the liquid often turned out three times more damaging than the effect of the pump. Such monitoring fluid flow controlling means usually has more negative impact than the influence of the pump.

Rapid changes in the transition pressures can lead to fluctuations or resonance. It can damage the pipe, resulting in leakage or rupture. Experiments show that the flow in the pipe will be uniform if its fluctuations are very small, say, located at 24 (*KM/h*).

This corresponds to about 0.45% of the velocity of pressure. In this case, the flow fluctuations can be easily accumulated and redeemed until the next perturbation. Fluctuations of the flow are no fluctuations of pressure.

3.2   BASIC EQUATIONS DESCRIBING THE SPHERICALLY SYMMETRIC MOTION OF A BUBBLE BINARY SOLUTION

The dynamics and heat and mass transfer of vapor bubble in a binary solution of liquids, in [8], was studied for significant thermal, diffusion and inertial effect. It was assumed that binary mixture with a density ρ_l, consisting of components 1 and 2, respectively, the density ρ_l and ρ_2.

Moreover:

$$\rho_1 + \rho_2 = \rho_l,$$

where, the mass concentration of component one of the mixture [19-20] also consider a two temperature model of interphase heat exchange for the bubble liquid. This model assumes homogeneity of the temperature in phases [21, 22].

The intensity of heat transfer for one of the dispersed particles with an endless stream of carrier phase will be set by the dimensionless parameter of Nusselt Nu_l.

Bubble dynamics described by the Rayleigh equation:

$$R\overset{\bullet}{w_l} + \frac{3}{2}w_l^2 = \frac{p_1 + p_2 - p_\infty - 2\sigma/R}{\rho_l} - 4v_1\frac{w_l}{R} \tag{3.2.1}$$

where, p_1 and p_2 = the pressure component of vapor in the bubble, p_∞ = The pressure of the liquid away from the bubble, σ and v_1 = surface tension coefficient of kinematic viscosity for the liquid. Consider the condition of mass conservation at the interface [23].

Mass flow j_i^{TH} component $(i = 1,2)$ of the interface $r = R(t)$ in j_i^{TH} phase per unit area and per unit of time and characterizes the intensity of the phase transition is given by [24]:

$$j_i = \rho_i\left(\overset{\bullet}{R} - w_l - w_i\right), (i = 1,2) \tag{3.2.2}$$

where w_i = The diffusion velocity component on the surface of the bubble. The relative motion of the components of the solution near the interface is determined by Fick's law [25, 26].

$$\rho_1 w_1 = -\rho_2 w_2 = -\rho_l D \frac{\partial k}{\partial r}\bigg|_R \qquad (3.2.3)$$

If we add Equation (3.2.2), while considering that: $\rho_1 + \rho_2 = \rho_l$ and draw the Equation (3.2.3), we obtain [27, 28]

$$\dot{R} = w_l + \frac{j_1 + j_2}{\rho_l} \qquad (3.2.4)$$

Multiplying the first Equation (3.2.2) on ρ_2, the second in ρ_1 and subtract the second equation from the first. In view of (3.2.3) we obtain:

$$k_R j_2 - (1 - k_R) j_1 = -\rho_l D \frac{\partial k}{\partial r}\bigg|_R$$

Here k_R = the concentration of the first component at the interface.

With the assumption of homogeneity of parameters inside the bubble changes in the mass of each component due to phase transformations can be written as:

$$\frac{d}{dt}\left(\frac{4}{3}\pi R^3 \rho_i'\right) = 4\pi R^2 j_i \text{ or } \frac{R}{3}\dot{\rho_i'} + \dot{R}\rho_i' = j_i, \ (i = 1,2) \qquad (3.2.5)$$

Express the composition of a binary mixture in mole fractions of the component relative to the total amount of substance in liquid phase:

$$N = \frac{n_1}{n_1 + n_2} \qquad (3.2.6)$$

The number of moles i^{TH} component n_i, which occupies the volume V, expressed in terms of its density:

$$n_i = \frac{\rho_i V}{\mu_i} \qquad (3.2.7)$$

Substituting (3.2.7) in (3.2.6), we obtain:

$$N_1(k) = \frac{\mu_2 k}{\mu_2 k + \mu_1(1 - k)} \qquad (3.2.8)$$

By law, Raul partial pressure of the component above the solution is proportional to its molar fraction in the liquid phase, that is:

$$p_1 = p_{S1}(T_v) N_1(k_R) \text{ and } p_2 = p_{S2}(T_v)[1 - N_1(k_R)] \qquad (3.2.9)$$

Equations of state phases have the form:

$$p_i = BT_v \rho_i' / \mu_i, \ (i = 1,2), \qquad (3.2.10)$$

where, B = Gas constant,

T_v = The temperature of steam,

ρ_i' = The density of the mixture components in the vapor bubble,

μ_i = Molecular weight,

p_{si} = Saturation pressure.

The boundary conditions $r = \infty$ and on a moving boundary can be written as:

$$k\Big|_{r=\infty} = k_0, k\Big|_{r=R} = k_R, \ T_l\Big|_{r=\infty} = T_0, T_l\Big|_{r=R} = T_v \tag{3.2.11}$$

$$j_1 l_1 + j_2 l_2 = \lambda_l D \frac{\partial T_l}{\partial r}\Big|_{r=R} \tag{3.2.12}$$

where, l_i = specific heat of vaporization [29,30].

By the definition of Nusselt parameter the dimensionless parameter characterizing the ratio of particle size and the thickness of thermal boundary layer in the phase around the phase boundary and determined from additional considerations or experience [31, 32].

The heat of the bubble's intensity with the flow of the carrier phase will be further specified as:

$$\left(\lambda_l \frac{\partial T_l}{\partial r}\right)_{r=R} = Nu_l \cdot \frac{\lambda_l (T_0 - T_v)}{2R} \tag{3.2.13}$$

In [33] obtained an analytical expression for the Nusselt parameter:

$$Nu_l = 2\sqrt{\frac{\omega R_0^2}{a_l}} = 2\sqrt{\frac{R_0}{a_l}\sqrt{\frac{3\gamma p_0}{\rho_l}}} = 2\sqrt{3\gamma \cdot Pe_l} \tag{3.2.14}$$

where, $a_l = \dfrac{\lambda_l}{\rho_l c_l}$ = thermal diffusivity of fluid,

$$Pe_l = \frac{R_0}{a_l}\sqrt{\frac{p_0}{\rho_l}} = \text{Peclet number.}$$

The intensity of mass transfer of the bubble with the flow of the carrier phase will continue to ask by using the dimensionless parameter Sherwood Sh:

$$\left(D \frac{\partial k}{\partial r}\right)_{r=R} = Sh \cdot \frac{D(k_0 - k_R)}{2R}$$

here, D = Diffusion coefficient,

k = The concentration of dissolved gas in liquid,

The subscripts 0 and R refer to the parameters in an undisturbed state and at the interface.

We define a parameter in the form of Sherwood [33]:

$$Sh = 2\sqrt{\frac{aR_0^2}{D}} = 2\sqrt{\frac{R_0}{D}\sqrt{\frac{3\gamma p_0}{\rho_l}}} = 2\sqrt{\sqrt{3\gamma} \cdot Pe_D} \qquad (3.2.15)$$

where, $Pe_D = \dfrac{R_0}{D}\sqrt{\dfrac{p_0}{\rho_l}}$ = diffusion Peclet number.

The system of Equations (3.2.1)–(3.2.15) is a closed system of equations describing the dynamics and heat transfer of insoluble gas bubbles with liquid.

3.3 THE BRAKING EFFECT OF THE INTENSITY OF PHASE TRANSFORMATIONS IN BOILING BINARY SOLUTIONS

If we use (3.2.7)–(3.2.9), we obtain relations for the initial concentration of component 1:

$$k_0 = \frac{1-\chi_2^0}{1-\chi_2^0 + \mu(\chi_1^0 - 1)}, \ \mu = \mu_2/\mu_1, \ \chi_i^0 = p_{si0}/p_0, \ i = 1,2 \qquad (3.3.1)$$

where, μ_2, μ_1 = Molecular weight of the liquid components of the mixture, p_{si0} = saturated vapor pressure of the components of the mixture at an initial temperature of the mixture T_0, which were determined by integrating the Clausius Clapeyron relation. The parameter χ_i^0 is equal to:

$$\chi_i^0 = \exp\left[\frac{l_i\mu_i}{B}\left(\frac{1}{T_{ki}} - \frac{1}{T_0}\right)\right], \qquad (3.3.2)$$

Gas phase liquid components in the derivation of (3.3.2) seemed perfect gas equations of state:

$$p_i = \rho_i B T_i / \mu_i.$$

here, B = Universal gas constant,

p_i = The vapor pressure inside the bubble T_i to the temperature in the ratio of (3.3.2)

T_{ki} = Temperature evaporating the liquid components of binary solution at an initial pressure p_0,

l_i = Specific heat of vaporization.

The initial concentration of the vapor pressure of component P_0 is determined from the relation:

$$c_0 = \frac{k_0 \chi_1^0}{k_0 \chi_1^0 + (1 - k_0) \chi_2^0} \qquad (3.3.3)$$

The problem of radial motions of a vapor bubble in binary solution [34] was solved. It was investigated at various pressure drops in the liquid for different initial radii R_0 for a bubble. It is of great practical interest of aqueous solutions of ethanol and ethylene glycol.

It was revealed an interesting effect. The parameters characterized the dynamics of bubbles in aqueous ethyl alcohol. It was studied in the field of variable pressure lie between the limiting values of the parameter P_0 for pure components.

The pressure drops and consequently the role of diffusion are assumed unimportant. The pressure drop along with the heat dissipation is included diffusion dissipation. The rate of growth and collapse of the bubble is much higher than in the corresponding pure components of the solution under the same conditions. A completely different situation existed during the growth and collapse of vapor bubble in aqueous solutions of ethylene glycol.

In this case, the effect of diffusion resistance, leaded to inhibition of the rate of phase transformations. The growth rate and the collapse of the bubble is much smaller than the corresponding values but for the pure components of the solution. Further research and calculations have to give a physical explanation for the observed effect. In [35–38] studied the influence of heat transfer and diffusion on damping of free oscillations of a vapor bubble binary solution.

It was found that the dependence of the damping rate of oscillations of a bubble of water solutions of ethanol, methanol, and toluene monotonic on k_0.

It was mentioned for the aqueous solution of ethylene glycol similar dependence with a characteristic minimum at:

$$k_0 \approx 0,02$$

Moreover, for $0,01 \le k_0 \le 0,3$ decrement, binary solution has less damping rates for pulsations of a bubble in pure (one-component) water and ethylene glycol.

This means that in the range of concentrations of water:

$$0,01 \le k_0 \le 0,3$$

Pulsations of the bubble (for water solution of ethylene glycol) decay much more slowly and there is inhibition of the process of phase transformations. A similar process was revealed and forced oscillations of bubbles in an acoustic field [35].

The influence of non-stationary heat and mass transfer [39] processes was investigated in the propagation of waves in a binary solution of liquids with bubbles. The influence of component composition and concentration of binary solution was investigated on the dispersion, dissipation, and attenuation of monochromatic waves in two phase, two component media.

The aqueous solution of ethyl alcohol in aqueous ethylene glycol decrements showed perturbations less relevant characteristics of pure components of the solution.

The unsteady interphase heat transfer revealed in [40] calculation, the structure of stationary shock waves in bubbly binary solutions. The problem signifies on effect a violation of monotonicity behavior of the calculated curves for concentration, indicating the presence of diffusion resistance.

In some of binary mixtures, it is seen the effect of diffusion resistance. It is led to inhibition of the intensity of phase transformations.

The physical explanation revealed the reason for an aqueous solution of ethylene glycol. The pronounced effect of diffusion resistance is related to the solution with limited ability. It diffuses through the components of $D=10^{-9}$ (m^2/sec).

D = Diffusion coefficient volatility of the components is very different, and thus greatly different concentrations of the components in the solution and vapor phase.

In the case of aqueous solution of ethanol volatility component are roughly the same $\chi_1^0 \approx \chi_2^0$

In accordance with (3) $c_0 \approx k_0$, so the finiteness of the diffusion coefficient does not lead to significant effects in violation of the thermal and mechanical equilibrium phases.

Figure 4 and Figure 5 show the dependence $k_0(c_0)$ of ethyl alcohol and ethylene glycol's aqueous solutions. From Figure 4 it is clear that almost the entire range of k_0, $k_0 \approx c_0$.

At the same time for an aqueous solution of ethylene glycol, by the calculations and Figure 4 $0.01 \le k_0 \le 0.3$ $k_0 \le c_0$,, and when $k_0 > 0.3$ $k_0 \sim c_0$.

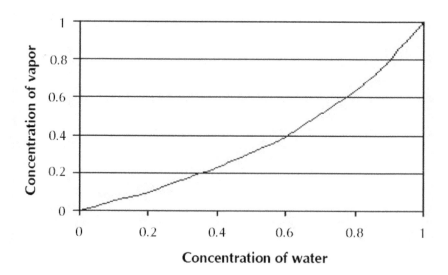

FIGURE 4   The dependence ($k_0(c_0)$) for an aqueous solution of ethanol.

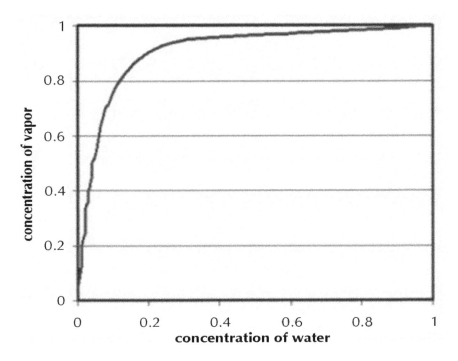

FIGURE 5   Dependence of ($k_0(c_0)$) for an aqueous solution of ethylene glycol.

In Figure 6 and Figure 7 show the boiling point of the concentration for the solution of two systems.

when, $k_0 = 1, c_0 = 1$ and get clean water to steam bubbles. It is for boiling of a liquid at

$T_0 = 373^0 K$.

If $k_0 = 0$, $c_0 = 0$ and have correspondingly pure bubble ethanol $(T_0 = 350^0 K)$ and ethylene glycol $(T_0 = 470^0 K)$.

It should be noted that, all works regardless of the problems in the mathematical description of the cardinal effects of component composition of the solution shows the value of the parameter β equal:

$$\beta = \left(1 - \frac{1}{\gamma}\right) \frac{(c_0 - k_0)(N_{c_0} - N_{k_0})}{k_0(1 - k_0)} \frac{c_l}{c_{pv}} \left(\frac{c_{pv} T_0}{L}\right)^2 \sqrt{\frac{a_l}{D}} \qquad (3.3.4)$$

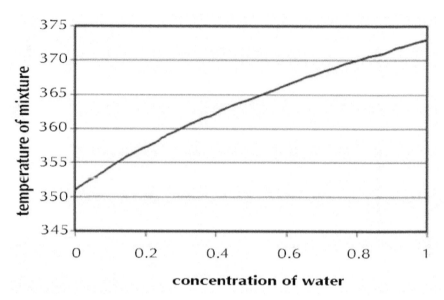

FIGURE 6   The dependence of the boiling temperature of the concentration of the solution to an aqueous solution of ethanol.

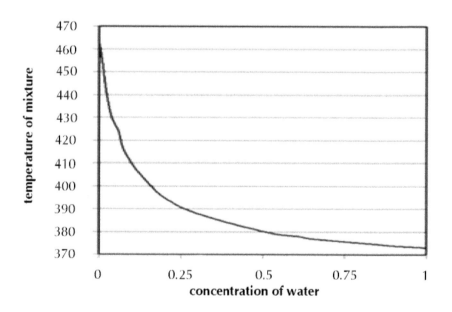

FIGURE 7   The dependence of the boiling point of the concentration of the solution to an aqueous solution of ethylene glycol.

where, N_{k_0}, N_{c_0} = Molar concentration of 1th component in the liquid and steam

$$N_{k_0} = \frac{\mu k_0}{\mu k_0 + 1 - k_0},$$

$$N_{c_0} = \frac{\mu c_0}{\mu c_0 + 1 - c_0} \quad \gamma = \text{Adiabatic index},$$

c_l and $c_{p,}$ respectively the specific heats of liquid and vapor at constant pressure,
a_l = Thermal diffusivity,
$L = l_1 c_0 + l_2 (1 - c_0)$

We also note that option (4) is a self-similar solution describing the growth of a bubble in a superheated solution. This solution has the form [41]:

$$R = 2\sqrt{\frac{3}{\pi} \frac{\lambda_l \Delta T \sqrt{t}}{L \rho_v \sqrt{a_l (1 + \beta)}}} \tag{3.3.5}$$

here, ρ_v = Vapor density,
t = Time
R = Radius of the bubble,
λ_l = the coefficient of thermal conductivity,
ΔT = Overheating of the liquid.

Figure 8 and Figure 9 shows the dependence $\beta(k_0)$ for the binary solutions. For aqueous ethanol β is negative for any value of concentration and dependence on k_0 is monotonic.

For an aqueous solution of ethylene glycol β = is positive and has a pronounced maximum at $k_0 = 0,2$.

As a result of present work at low pressure drops respectively superheating and super cooling of the liquid), diffusion does not occur in aqueous solutions of ethyl alcohol. By approximate equality of k_0 and c_0 all calculated dependence lie between the limiting curves for the case of one component constituents of the solution.

They are included dependence of pressure, temperature, vapor bubble radius, the intensity of phase transformations, and so from time to time. The pressure difference becomes important diffusion processes. Mass transfer between bubble and liquid is in a more intensive mode than in single component constituents of the solution. In particular, the growth rate of the bubble in a superheated solution is higher than in pure water and ethyl alcohol. It is because of the negative β according to (3.3.5).

In an aqueous solution of ethylene glycol, there is the same perturbations due to significant differences between k_0 and c_0. It is especially when $0,01 \le k_0 \le 0,3$, the effect of diffusion inhibition contributes to a significant intensity of mass transfer. In particular, during the growth of the bubble, the rate of growth in solution is much lower than in pure water and ethylene glycol. It is because of the positive β by (3.3.5).

Moreover, the maximum braking effect is achieved at the maximum value of β, when $k_0 = 0,2$. A similar pattern is observed at the pulsations and the collapse of the bubble. The dependence of the damping rate of fluctuations in an aqueous solution of ethyl alcohol from the water concentration is monotonic is shown in [35]. Aqueous solution of ethylene glycol dependence of the damping rate has a minimum at $k_0 = 0,2$, $0,01 \leq k_0 \leq 0,3$.

The function decrement is small respectively large difference between k_0 and c_0 and β takes a large value. These ranges of concentrations in the solution have significant effect of diffusion inhibition. For aqueous solutions of glycerin, methanol, toluene, and so on, calculations are performed. The comparison with experimental data confirms the possibility of theoretical prediction of the braking of heat and mass transfer.

It was analyzed the dependence of the parameter β, decrement of oscillations of a bubble from the equilibrium concentration of the mixture components. Therefore, in every solutions, it was determined the concentration of the components of a binary mixture.

concentration of water

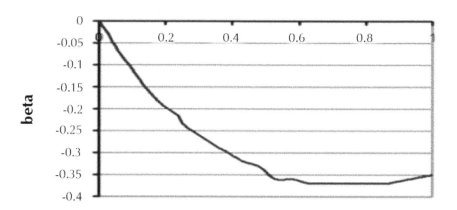

FIGURE 8   The dependence $\beta(k_0)$ for an aqueous solution of ethanol.

The Figure 10 and Figure 11 are illustrated by theoretical calculations. These figures defined on the example of aqueous solutions of ethyl alcohol and ethylene glycol (antifreeze used in car radiators). It is evident that the first solution is not suitable to the task.

The aqueous solution of ethylene glycol with a certain concentration is theoretically much more slowly boils over with clean water and ethylene glycol. This confirms the reliability of the method.

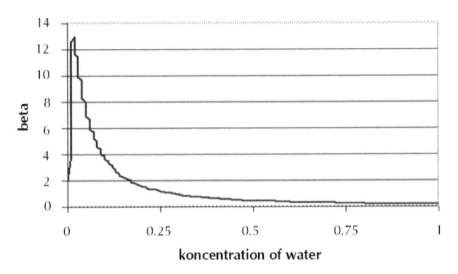

FIGURE 9   Dependence of $\beta(k_0)$ for an aqueous solution of ethylene glycol.

The calculations shows that solution is never freezes. The same method can offer concrete solutions for cooling of hot parts and components of various machines and mechanisms.

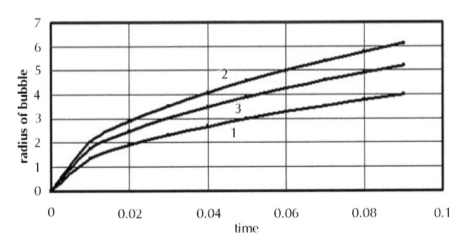

FIGURE 10   Dependence from time of vapor bubble radius. 1—water, 2—ethyl spirit, 3— water mixtures of ethyl spirit.

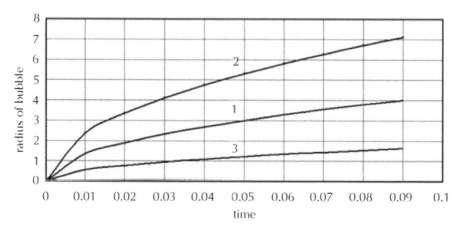

FIGURE 11   Dependence from time of vapor bubble radius. 1—water, 2—ethylene glycol, 3—water mixtures of ethylene glycol.

The solution of the reduced system of equations revealed an interesting effect. The parameters were characterized the dynamics of bubbles in aqueous ethyl alcohol in the field of variable pressure. They lied between the limiting values of relevant parameters for the pure components. It was for the case which pressure drops and consequently the role of diffusion was unimportant.

A completely different situation is observed during the growth and collapse of vapor bubble in aqueous solutions of ethylene glycol. The effect of diffusion resistance, leads to inhibition of the rate of phase transformations. For pure components of the solution, the growth and the collapse rate of the bubble is much smaller than the corresponding values.

3.4   THE STRUCTURE OF THE PRESSURE WAVE FOR A SIMPLE PIPELINE SYSTEM

The wave dynamics of dispersed two phase mixtures, in contrast to homogeneous media, is determined by processes of interaction between phases. The essential difference between mechanical and physical-chemical properties of the phases leads to the fact that the external disturbance has on the carrier and the dispersed phase of different actions. As a result of the wave front of finite intensity phase is no longer in equilibrium and the resulting relaxation processes can significantly influence the course.

The rates of relaxation processes determine the structure of individual relaxation zones for elementary waves, and the whole flow as a whole.

For example, the rapid mechanical fragmentation of drops for shock and detonation waves leads to the formation of a large number of secondary droplets, the surface area that is several orders of magnitude higher than those for the source of aerosol.

That leads to rapid evaporation of the liquid, mixing the vapor on gaseous oxidizer and the formation of a homogeneous air-fuel mixture [42]. The basis for studying the dynamics of disperses mixtures is a description of the mechanical motion of the mixture "in general" under review at the scale of the whole problem. Extensive interactions of the phases (including those caused by the deformation, fragmentation, evaporation and mixing), going at the scale of individual particles, determined by this macro scale motion, providing, in turn, a significant inverse effect on him.

Currently, conventional mathematical hydrodynamics model of heterogeneous environments is a model multispeed continuum, the most complete exposition in the works [43].

Investigate the structure of stationary shock waves in binary mixtures, propagating with the velocity of pressure wave. This speed is a fundamental parameter for modeling hydraulic transients. Consider the effect of the dynamics of vapor bubbles in gas-liquid two component mixtures on the propagation of shock waves, the effect of unsteady forces on transient flows, division of water columns in the vapor or steam bubbles, control steam pressure in cavities to pump up the direction of the pipeline. Solving this problem is extremely important to study all the conditions under which the piping system having adverse transients, in particular in the pumps and valves. In addition, it is important to develop methods of protection and devices to be used during the design and construction of separate parts of the system, as well as to identify their practical shortcomings.

In solving problems for the management of transients caused by abrupt change in pressure, suggests two possible strategies. The first strategy is to minimize the possibility of transient conditions during the design, determine the appropriate methods of flow control to eliminate the possibility of emergency and unusual situations in the system. The second strategy is to provide for the establishment of security device to control the possible transients due to events beyond control, such as equipment failures and power supply [44-45].

The Figure 12 and 13 shows the curves of pressure of time, received both experimental and theoretical methods. It is seen that the results of theoretical calculations agree well with experimental data [46].

In the experiments, the pipeline was equipped with a valve at the end of the main pipes, joined by timers to record time of closing. The characteristics of water hydraulic impact measured and recorded by extensometers in the computer memory. Pumping of water to the system was carried out by the pool, which allowed the pressure to stabilize the inlet.

The experiments were performed for three cases:

1. A simple positive hydraulic shock for a straight pipe of constant diameter. The measured characteristics were the basis for assessing the impact of changes in diameter and the local diversion to the distribution of water hammer.
2. Water hammer in pipelines with diameters change: contraction and expansion.
3. Water hammer in pipelines with a local diversion in two scenarios: the outflow from the brain to super compression pool and a free outflow from the brain (to atmospheric pressure, with the ability to absorb air in the negative phase). This was the main reason for the air intake in the negative phase.

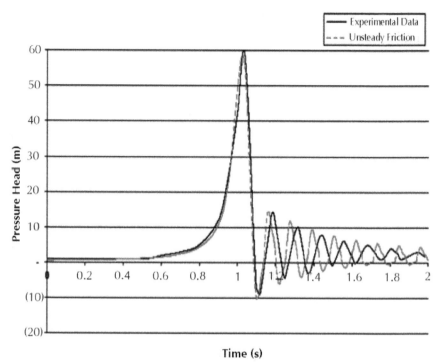

FIGURE 12 The structure of the pressure wave for a simple pipeline system using the model of equilibrium and no equilibrium friction.

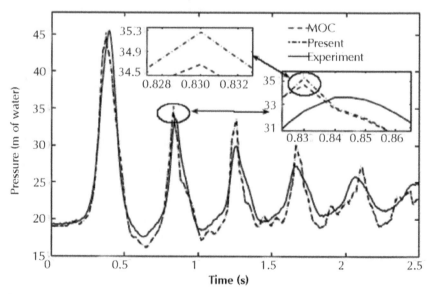

FIGURE 13 Comparison the results with experimental data.

Separation columns for pump power off on pipeline were carried out for two cases:

(a) With the surge tank and the assumption of local diversion: In this case, the air was sucked into the pipeline,

(b) Without the surge tank, provided the local diversion.

It can be recorded pressure wave and the wave velocity in fast transients up to 5 ms (in this study to 1 s). The assessment procedure was used to analyze Curve data, which were obtained on real systems. Software regression curve providing customized features and provided a regression analysis. Thus, the regression model (first model) was used in the final procedure. This model was compared with the results of the model performance (second model). The calculation results allowed us to develop a technical solution for the management of transition processes in the pipeline system [47]. Figure 14 shows a scheme of how the developed device.

The surge tank with double bottom manages transitions, converting accumulated potential energy of water to kinetic energy. In critical situations during periods of rapid change in the nature of the flow of water from the reservoir flows into the system piping. The tank is usually located in the pumping station or highest point in the water system profile. In the vertical jumpers reservoir can be drilled various holes to control the supply of water from the system into the reservoir. But this design results in very little loss of flux, spilling from the tank. If there is overflow and leak, the tank can also act as a protective device against unintentional pressure increase.

Thus, the device may serve as a way of protection from the overflow. There is another important problem. The problem is related to stability in the reservoir. It is necessary for avoiding rapid rising or lowering the water level in the reservoir.

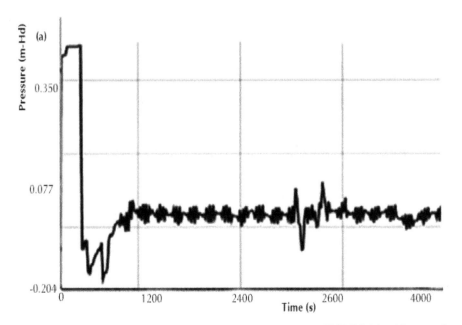

FIGURE 14   *(Continued)*

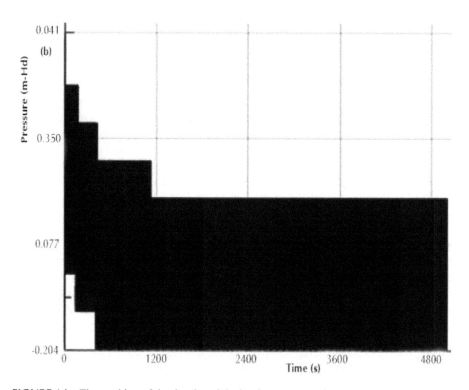

FIGURE 14   The working of the developed device for water supply.

Therefore, the surface area of the tank must be much greater cross section of the pipeline [48].

3.5   CONCLUSION

It was evaluated by a comprehensive approach the strength of pipeline systems consisting of technical diagnostics segment of the pipe and the mathematical modeling of the pipeline. Results implicated on prediction of the selected model.

It was used a mathematical model for determining the nature of pulse propagation of pressure in pipelines. It was concluded the presence of leaks, bends, and local resistance. The model provides a basis for developing a new method for determining leakage and unauthorized connections. It gives a clear idea about the nature of pulse propagation of pressure in pipelines.

The experimental setup and performed on studies confirmed the validity of the proposed mathematical model, and comparing the results of calculations and experiments showed satisfactory agreement.

Industrial testing method for detection of leaks and to combat water hammer for the water pipeline showed the effectiveness of the proposed methods.

It was confirmed the two temperature model theoretically on anomalous braking speed of phase transformations in boiling of binary mixtures.

It was investigated a comparison on accuracy of the numerical methods, regression model and the method of characteristics for the analysis of transient flows. It was shown that the method of characteristics is computationally more efficient for the analysis of large water pipelines.

KEYWORDS

- **Bubble binary solution**
- **Hydraulic transition**
- **Nusselt parameter**
- **Vapor bubble**
- **Vena contraction**
- **Water hammer**

REFERENCES

1. Qu, W., Mala, G. M., and Li, D. *Heat Transfer for Water Flow in Trapezoidal Silicon Microchannels*, pp. 399–404 (1993).
2. Hariri Asli, K., Nagiyev, F. B., Haghi, A. K., and Aliyev, S. A. A computational approach to study fluid movement, 1st Festival on Water and Wastewater Research and Technology, Tehran, Iran, pp. 27–32, http://isrc.nww.co.ir (December 12–17, 2009).
3. Peng, X. F.and Peterson, G. P. Convective Heat Transfer and Flow Friction for Water Flow in Microchannel Structure, *Int. J. Heat Mass Transfer* **36** 2599–2608 (1996).
4. Bergant Anton. *Discrete Vapor Cavity Model with Improved Timing of Opening and Collapse of Cavities*, pp. 1–11 (1980),
5. Ishii, M. *Thermo-Fluid Dynamic Theory of Two-Phase Flow.* Collection de D. R. Liles and W. H. Reed (Eds.). A Sern-Implict Method for Two-Phase Fluid la Direction des Etudes et. Recherché d'Electricite de France, Vol. 22 Dynamics. Journal of Computational Physics 26, Paris, pp. 390–407 (1975).
6. Hariri Asli, K., Nagiyev, F. B., Haghi, A. K. and Aliyev, S. A. *A computational approach to study fluid movement.* 1st Festival on Water and Wastewater Research and Technology, Tehran, Iran, pp. 27–32, http://isrc.nww.co.ir (December 12–17, 2009).
7. Pickford, J. *Analysis of Surge*, Macmillian, London, pp. 153–156 (1969),
8. *Pipeline Design for Water and Wastewater.* American Society of Civil Engineers, New York, p. 54 (1975).
9. Xu, B., Ooi, K. T., Mavriplis, C. and Zaghloul, M. E. *Viscous dissipation effects for liquid flow in microchannels. Micorsystems*, pp. 53–57 (2002),
10. Fedorov, A. G. and Viskanta, R. Three-dimensional Conjugate Heat Transfer into Microchannel Heat Sink for Electronic Packaging. *Int. J. Heat Mass Transfer*, **43**, 399–415 (2000).
11. Tuckerman, D. B. Heat transfer microstructures for integrated circuits. Ph.D. thesis, Stanford University, pp. 10–120 (1984,)
12. Harms, T. M., Kazmierczak, M. J., Cerner, F. M., Holke, A., Henderson, H. T., Pilchowski, H. T. and Baker, K. Experimental Investigation of Heat Transfer and Pressure Drop through Deep Micro channels in a (100) Silicon Substrate, in Proceedings of the ASME. *Heat Transfer Division, HTD*, **351**, 347–357 (1997),
13. Holland, F. A. and Bragg, R. *Fluid Flow for Chemical Engineers.* Edward Arnold Publishers, London, pp. 1–3 (1995),

14. Lee, T. S. and Pejovic, S. Air influence on similarity of hydraulic transients and vibrations. *ASME J. Fluid Eng*, **118**(4), 706–709.
15. Li, J. and McCorquodale, A., Modeling Mixed Flow in Storm Sewers, *Journal of Hydraulic Engineering, ASCE*, **125**(11) 1170–1180 (1999).
16. Minnaert, M. On musical air bubbles and the sounds of running water. *Phil. Mag.*, **16**(7) 235–248 (1933),
17. Moeng, C. H., McWilliams, J. C., Rotunno, R., Sullivan, P. P. and Weil, J. Investigating 2D modeling of atmospheric convection in the PBL. *J. Atm. Sci*, **61**, 889–903 (2004).
18. Tuckerman, D. B., Pease, R. F. W. High performance heat sinking for VLSI. *IEEE Electron device lette, DEL* **2**, 126–129 (1981).
19. Nagiyev, F. B. and Khabeev, N. S. Bubble dynamics of binary solutions. *High Temperature*, **27**(3) 528–533 (1988).
20. Shvarts, D., Oron, D., Kartoon, D., Rikanati, A., and Sadot, O. Scaling laws of nonlinear Rayleigh-Taylor and Richtmyer-Meshkov instabilities in two and three dimensions. *C. R. Acad. Sci. Paris IV*, **719** 312 (2000).
21. Cabot, W. H., Cook, A. W., Miller, P. L., Laney, D. E., Miller, M. C., and Childs, H. R. Large eddy simulation of Rayleigh-Taylor instability. *Phys. Fluids*, **17** 91–106 (September, 2005).
22. Cabot, W. *Lawrence Livermore National laboratory*. University of California, Livermore, CA, Physics of Fluids 94–550 (2006).
23. Goncharov, V. N. Analytical model of nonlinear, single-mode, classical Rayleigh-Taylor instability at arbitrary Atwood numbers. *Phys. Rev. Letters*, **88**, **134502** 10–15 (2002).
24. Ramaprabhu, P. and Andrews, M. J. Experimental investigation of Rayleigh-Taylor mixing at small Atwood numbers. *J. Fluid Mech.*, **502** 233 (2004).
25. Clark, T. T. A numerical study of the statistics of a two-dimensional Rayleigh-Taylor mixing layer. *Phys. Fluids*, **15** 2413 (2003).
26. Cook, A. W., Cabot, W., and Miller, P. L. The mixing transition in Rayleigh-Taylor instability. *J. Fluid Mech.*, **511** 333 (2004).
27. Waddell, J. T., Niederhaus, C. E., and Jacobs, J. W. Experimental study of Rayleigh-Taylor instability: Low Atwood number liquid systems with single-mode initial perturbations. *Phys. Fluids*, **13** 1263–1273 (2001).
28. Weber, S. V., Dimonte, G., and Marinak, M. M. Arbitrary Lagrange-Eulerian code simulations of turbulent Rayleigh-Taylor instability in two and three dimensions. *Laser and Particle Beams*, **21** 455 (2003).
29. Dimonte, G., Youngs, D., Dimits, A., Weber, S., and Marinak, M. A comparative study of the Rayleigh-Taylor instability using high-resolution three-dimensional numerical simulations: the Alpha group collaboration. *Phys. Fluids*, **16** 1668 (2004).
30. Young, Y. N., Tufo, H., Dubey, A., and Rosner, R. On the miscible Rayleigh-Taylor instability: two and three dimensions. *J. Fluid Mech.*, **447**, **377** 2003–2500 (2001).
31. George, E. and Glimm, J. Self-similarity of Rayleigh-Taylor mixing rates. *Phys. Fluids*, **17**, **054101** 1–3 (2005).
32. Oron, D., Arazi, L., Kartoon, D., Rikanati, A., Alon, U., and Shvarts, D. Dimensionality dependence of the Rayleigh-Taylor and Richtmyer-Meshkov instability late-time scaling laws. *Phys. Plasmas*, **8** 2883 (2001).
33. Nigmatulin, R. I., Nagiyev, F. B., and Khabeev, N. S. *Effective heat transfer coefficients of the bubbles in the liquid radial pulse*. Mater. Second-Union. Conf. Heat Mass Transfer. Heat massoob-men in the biphasic with, Minsk, Vol. 5, pp. 111–115 (1980).
34. Nagiyev, F. B. and Khabeev, N. S. Bubble dynamics of binary solutions. *High Temperature*, **27**(3) 528–533 (1988).

35. Nagiyev, F. B. *Damping of the oscillations of bubbles boiling binary solutions*. Mater. VIII Resp. Conf. mathematics and mechanics. Baku, pp. 177–178 (October 26–29, 1988).
36. Nagiyev, F. B. and Kadyrov, B. A. Small oscillations of the bubbles in a binary mixture in the acoustic field. *Math. AN Az.SSR Ser. Physico-tech. and mate. Science*, (1) 23–26 (1986).
37. Nagiyev, F. B. *Dynamics, heat and mass transfer of vapor-gas bubbles in a two-component liquid*. Turkey-Azerbaijan petrol semin, Ankara, Turkey, pp. 32–40 (1993).
38. Nagiyev, F. B. *The method of creation effective coolness liquids, Third Baku international Congress*. Azerbaijan Republic, Baku, pp. 19–22 (1995).
39. Nagiyev, F. B. The linear theory of disturbances in binary liquids bubble solution. *Dep. In VINITI*, **86**(405) 76–79 (1986).
40. Nagiyev, F. B. Structure of stationary shock waves in boiling binary solutions. *Math. USSR, Fluid Dynamics*, (1) 81–87 (1989).
41. Rayleigh, L. On the pressure developed in a liquid during the collapse of a spherical cavity. *Philos. Mag. Ser.*, **6, 34**(200) 94–98 (1917).
42. Perry, R. H., Green, D. W., and Maloney, J. O. *Perry's Chemical Engineers Handbook, 7th Edition*. McGraw-Hill, New York, pp. 1–61 (1997).
43. Nigmatulin, R. I. *Dynamics of multiphase media*, Nauka, Moscow, Vol. 1, 2, pp. 12–14 (1987).
44. Kodura, A. and Weinerowska, K. The influence of the local pipeline leak on water hammer properties. *Materials of the II Polish Congress of Environmental Engineering*. Lublin, pp. 125–133 (2005).
45. Kane, J., Arnett, D., Remington, B. A., Glendinning, S. G., and Baz´an, G. Two-dimensional versus three-dimensional supernova hydrodynamic instability growth. *Astrophys. J.*, **528** 989–994 (2000).
46. Quick, R. S. Comparison and Limitations of Various Water hammer Theories. *J. of Hyd. Div., ASME* 43–45 (May, 1933).
47. Jaeger, C. *Fluid Transients in Hydro-Electric Engineering Practice*. Blackie & Son Ltd., pp. 87–88 (1977).
48. Jaime Suárez, A. *Generalized water hammer algorithm for piping systems with unsteady friction*, pp. 72–77 (2005).

4 Water Hammer and Surge Wave Modeling

CONTENTS

4.1 INTRODUCTION

This chapter discuss on Eulerian based model for water hammer. This model was defined by the method of characteristics (MOC), finite difference form. The method was encoded into an existing hydraulic simulation model. The surge wave was assumed as a failure factor in an elastic case of water pipeline with free water bubble. The results were compared by regression analysis. It indicated that the accuracy of the Eulerian based model for water transmission line.

4.2 WATER HAMMER AS A FLUID DYNAMICS PHENOMENON

Water hammer as a fluid dynamics phenomenon is an important case study for designer engineers. This phenomenon has a complex mathematical behavior. Today high technology provides a suitable condition for finding new methods in order to reduction of water hammer disaster. Water hammer disaster can be happened at earthquake or tsunami. For example, at these critical conditions, water transmission pipeline control at power plants, water treatment plants, water transmission and distribution plants will be at high risk due to damage or failure hazard. As a side effect of water hammer phenomenon, this situation increases the probability of surge wave generation. Surge pressure and velocity of surge wave acts at fast transients, down to 5 ms. So it must be detected on actual systems (by field tests) by high technology and high speed detectors. Also, besides the flow and pressure, it must be computed by computational model. The recording of fast transients needs to use the high technology and online data intercommunication. Water transmission failure sometimes happens due to unusual factors which can suddenly change in the boundaries of the system. High surge pressure at earth quake, pump power off and inlet air by the air inlet valve, high discharge rate due to connections and consumers are some of the unusual factors which suddenly change in the boundaries of the system. Most of the transients in water and wastewater systems are the result of changes in the properties and boundaries of the system [1]. It can generate the spread of the surge wave and changes at liquids properties in pipes and channels. It causes the formation and collapse of vapor bubbles or cavitations and air leakage Hariri Asli, et al. [2]. The study of hydraulic transients began with the work of Zhukovsky [1]. Many researchers have made significant contributions in this area, including Parmakian [3], Wood [4] who popularized and perfected the graphical method of calculation. Wylie and Streeter [5] MOC combined with computer modeling. Subject of transients in liquids are still growing fast around the world. Brunone et al. [6], Koelle and Luvizotto [7], Filion and Karney [8], Hamam and Mc Corquodale [9], Savic and Walters [10], Walski and Lutes [11], Lee and Pejovic [12], have been developed various methods of investigation of transient pipe flow. These ranges of methods are included by approximate equations to numerical solutions of the nonlinear Navier–Stokes equations. Various methods have been developed to solve transient flow in pipes. These ranges have been formed from approximate equations to numerical solutions of the non-linear Navier–Stokes equations. Elastic theory describes the unsteady flow of a compressible liquid in an elastic system. Transient theory stem from the two governing equations. The continuity equation and the momentum equation are needed to determine velocity and surge pressure in a one-dimensional flow system. Solving these two equations produces a theoretical result that usually corresponds quite closely to actual system measurements, if the data and assumptions used to build the numerical model are valid. Among the approaches proposed to solve.

The single-phase (pure liquid) water hammer equations are the MOC finite differences (FD), wave characteristic method (WCM), finite elements (FE), and finite volume (FV). One difficulty that commonly arises relates to the selection of an appropriate level of time step to use for the analysis. The obvious trade off is between computational speed and accuracy. In general, for the smaller the time step, there is the longer the run time but the greater the numerical accuracy.

4.3 NUMERICAL SOLUTIONS OF THE NON-LINEAR NAVIER–STOKES EQUATIONS

An evaluation of surge or pressure wave in an elastic case with the free water bubble. It started with the solving of approximate equations by numerical solutions of the nonlinear Navier–Stokes equations based on the MOC. Then it derived the Zhukousky formula and velocity of surge or pressure wave in an elastic case with the high value of free water bubble. So the numerical modeling and simulation which was defined by "MOC" provided a set of results. Basically the "MOC" approach transforms the water hammer partial differential equations into the ordinary differential equations along the characteristic lines.

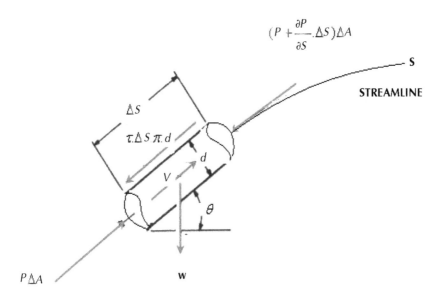

FIGURE 1   Newton second law (conservation of momentum equation) for fluid element.

It is defined as the combination of momentum equation (Figure 1) and continuity equation (Figure 2) for determining the velocity and pressure in a one dimensional flow system. The solving of these equations produces a theoretical result that usually corresponds quite closely to actual system measurements.

$$P\Delta A - (P + \frac{\partial P}{\partial S}.\Delta S)\Delta A - W.\sin\theta - \tau.\Delta S.\pi.d = \frac{W}{g}.\frac{dV}{dt}$$

Both sides are divided by m and with assumption:

$$\frac{\partial Z}{\partial S} = +\sin\theta$$

$$-\frac{1}{\partial}.\frac{\partial P}{\partial S} - \frac{\partial Z}{\partial S} - \frac{4\tau}{\gamma D} = \frac{1}{g}.\frac{dV}{dt} \qquad (4.2.1)$$

$$\Delta A = \frac{\Pi.D^2}{4}$$

If fluid diameter assumed equal to pipe diameter then:

$$\frac{-1}{\gamma}\cdot\frac{\partial P}{\partial S} - \frac{\partial Z}{\partial S} - \frac{4\tau_\circ}{\gamma.D} \tag{4.2.2}$$

$$\tau_\circ = \frac{1}{8}\rho.f.V^2$$

$$-\frac{1}{\gamma}\cdot\frac{\partial P}{\partial S} - \frac{\partial Z}{\partial S} - \frac{f}{D}\frac{V^2}{2g} = \frac{1}{g}\cdot\frac{dV}{dt} \tag{4.2.3}$$

$$V^2 = V\,|V|,\ \ \frac{dV}{dt} + \frac{1}{\rho}\cdot\frac{\partial P}{\partial S} + g\frac{dZ}{dS} + \frac{f}{2D}V|V| = 0 \qquad \text{(Euler equation) (4.2.4)}$$

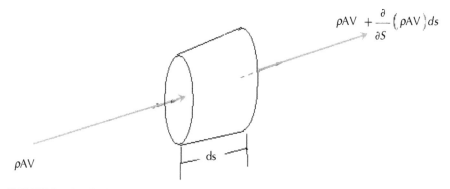

FIGURE 2   Continuity equation (conservation of mass) for fluid element.

For finding (V) and (P) we need to conservation of mass law (Figure 2)

$$\rho AV - \left[\rho AV - \frac{\partial}{\partial S}(\rho AV)dS\right] = \frac{\partial}{\partial t}(\rho AdS) - \frac{\partial}{\partial S}(\rho AV)dS = \frac{\partial}{\partial t}(\rho AdS) \tag{4.2.5}$$

$$-\left(\rho A\frac{\partial V}{\partial S}dS + \rho V\frac{\partial A}{\partial S}dS + AV\frac{\partial\rho}{\partial S}dS\right) = \rho A\frac{\partial}{\partial t}(dS) + \rho dS\frac{\partial A}{\partial t} + AdS\frac{\partial\rho}{\partial t}$$

$$\frac{1}{\rho}(\frac{\partial\rho}{\partial t} + V\frac{\partial\rho}{\partial S}) + \frac{1}{A}(\frac{\partial A}{\partial t} + V\frac{\partial A}{\partial S}) + \frac{1}{dS}\cdot\frac{\partial}{\partial t}(dS) + \frac{\partial V}{\partial S} = \circ \tag{4.2.6}$$

With $\ \dfrac{\partial\rho}{\partial t} + V\dfrac{\partial\rho}{\partial S} = \dfrac{d\rho}{dt}\ $ and $\ \dfrac{\partial A}{\partial t} + V\dfrac{\partial A}{\partial S} = \dfrac{dA}{dt}$

$$\frac{1}{\rho}\cdot\frac{d\rho}{dt} + \frac{1}{A}\cdot\frac{dA}{dt} + \frac{\partial V}{\partial S} + \frac{1}{dS}\cdot\frac{1}{dt}(dS) = \circ, \tag{4.2.7}$$

$$K = \left| \frac{d\rho}{\left(\frac{d\rho}{\rho} \right)} \right| \text{ (Fluid module of elasticity) then:}$$

$$\frac{1}{\rho} \cdot \frac{d\rho}{dt} = \frac{1}{k} \cdot \frac{d\rho}{dt} \tag{4.2.8}$$

Put (4.2.7) into (4.2.8)
Then:

$$\frac{\partial V}{\partial S} + \frac{1}{k} \cdot \frac{d\rho}{dt} + \frac{1}{A} \cdot \frac{dA}{dt} + \frac{1}{dS} \cdot \frac{d}{dt}(dS) = \circ$$

$$\rho \frac{\partial V}{\partial S} + \frac{d\rho}{dt} \rho \left[\frac{1}{k} + \frac{1}{A} \cdot \frac{dA}{d\rho} + \frac{1}{dS} \cdot \frac{d}{d\rho}(dS) \right] = \circ \tag{4.2.9}$$

$$\rho \left[\frac{1}{k} + \frac{1}{A} \cdot \frac{dA}{dt} + \frac{1}{dS} \cdot \frac{d}{d\rho}(dS) \right] = \frac{1}{C^2}$$

$$\text{Then } C^2 \frac{\partial V}{\partial S} + \frac{1}{\rho} \cdot \frac{d\rho}{dt} = \circ \qquad \text{(Continuity equation)} \tag{4.2.10}$$

Partial differential Equations (4.2.4) and (4.2.10) are solved by MOC:

$$\frac{dp}{dt} = \frac{\partial p}{\partial t} + \frac{\partial p}{\partial S} \cdot \frac{dS}{dt} \tag{4.2.11}$$

$$\frac{dV}{dt} = \frac{\partial V}{\partial t} + \frac{\partial V}{\partial S} \cdot \frac{dS}{dt} \tag{4.2.12}$$

Then

$$\left| \frac{\partial V}{\partial t} + \frac{1}{\rho} \frac{\partial p}{\partial S} + g \frac{dz}{dS} + \frac{f}{2D} V|V| = \circ, \right.$$

$$\left| C^2 \frac{\partial V}{\partial S} + \frac{1}{\rho} \frac{\partial P}{\partial t} = \circ, \right. \tag{4.2.13), (4.2.14}$$

By Linear combination of (4.2.13) and (4.2.14)

$$\lambda\left(\frac{\partial V}{\partial t}+\frac{1}{\rho}\frac{\partial p}{\partial S}+g.\frac{dz}{dS}+\frac{f}{2D}V|V|\right)+c^2\frac{\partial V}{\partial S}+\frac{1}{\rho}\frac{\partial p}{\partial t}=0 \qquad (4.2.15)$$

$$(\lambda\frac{\partial V}{\partial t}+C^2\frac{\partial V}{\partial S})+(\frac{1}{\rho}.\frac{\partial \rho}{\partial t}+\frac{\lambda}{\rho}.\frac{\partial P}{\partial S})+\lambda.g.\frac{dz}{dS}+\frac{\lambda.f}{2D}V|V|=0 \qquad (4.2.16)$$

$$\lambda\frac{\partial V}{\partial t}+C^2\frac{\partial V}{\partial S}=\lambda\frac{dV}{dt}\Rightarrow\lambda\frac{dS}{dt}=C^2 \qquad (4.2.17)$$

$$\frac{1}{\rho}.\frac{\partial p}{\partial t}+\frac{\lambda}{\rho}.\frac{\partial \rho}{\partial S}=\frac{1}{\rho}.\frac{d\rho}{dt}\Rightarrow$$
$$\frac{\lambda}{\rho}=\frac{1}{\rho}.\frac{dS}{dt} \qquad (4.2.18)$$

$$\left|\frac{C^2}{\lambda}\right.=\lambda\text{ (By removing }\frac{dS}{dt}\text{), }\lambda=\pm C$$

For $\lambda=\pm C$ From Equation (4.2.18) we have:

$$C\frac{dV}{dt}+\frac{1}{\rho}.\frac{dp}{dt}+C.g.\frac{dz}{dS}+C.\frac{f}{2D}V|V|=0$$

With dividing both sides by "C":

$$\frac{dV}{dt}+\frac{1}{c.\rho}\frac{dP}{dt}+g.\frac{dz}{dS}+\frac{f}{2D}V|V|=0 \qquad (4.2.19)$$

For $\lambda=-C$ by Equation (4.2.16):

$$\frac{dV}{dt}-\frac{1}{c.\rho}\frac{dp}{dt}+g\frac{dZ}{dS}+\frac{f}{2D}V|V|=0 \qquad (4.2.20)$$

If $\rho=\rho.g(H-Z)$

From Equations (4.2.9) and (4.2.10):

$$\left| \frac{dV}{dt} + \frac{g}{c} \cdot \frac{dH}{dt} + \frac{f}{2D}V|V| = 0 \right.$$
$$\left. if : \frac{dS}{dt} = C, \right.$$

(4.2.21) and (4.2.22)

$$\left| \frac{dV}{dt} + \frac{g}{c} \cdot \frac{dH}{dt} + \frac{f}{2D}V|V| = 0, \right.$$
$$\left. if : \frac{dS}{dt} = -C, \right.$$

(4.2.23) and (4.2.24)

The MOC is a finite difference technique which pressures were computed along the pipe for each time step (4.2.1)–(4.2.35).

The calculation automatically sub-divided the pipe into sections (intervals) and selected a time interval for computations Equations (4.2.22) and (4.2.24) are the characteristic Equation of (4.2.21) and (4.2.23)

If $f = 0$ Then Equation (4.2.23) will be (Figure 3)

$$\frac{dV}{dt} - \frac{g}{c} \cdot \frac{dH}{dt} = 0 \text{ or}$$

$$dH = \left(\frac{C}{g} \right) dV, (Zhukousky),$$

(4.2.25)

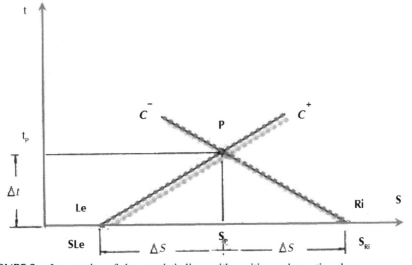

FIGURE 3 Intersection of characteristic lines with positive and negative slope.

If the pressure at the inlet of the pipe and along its length is equal to p_0, then slugging pressure undergoes a sharp increase:

$$\Delta p \,. \, p = p_0 + \Delta p$$

The Zhukousky formula is as flowing:

$$\Delta p = \left(\frac{C.\Delta V}{g} \right),$$

(4.2.26)

The speed of the shock wave is calculated by the formula:

$$C = \sqrt{\frac{g.\dfrac{E_W}{\rho}}{1 + \dfrac{d}{t_W} \cdot \dfrac{E_W}{E}}},$$

(4.2.27)

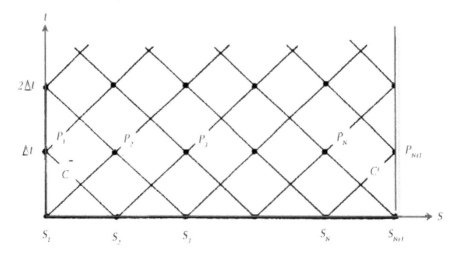

FIGURE 4   Set of characteristic lines intersection for assumed pipe.

By finite difference method of water hammer:

$$T_p - 0 = \Delta t$$

$$c^+: (V_p - V_{Le})/(T_P - \circ) + (\tfrac{g}{c})(H_p - H_{Le})/(T_P - \circ) + fV_{Le}|V_{Le}|/2D) = \circ, \qquad (4.2.28)$$

$$c^-: (V_p - V_{Ri})/(T_P - \circ) + (\tfrac{g}{c})(H_p - H_{Ri})/(T_P - \circ) + fV_{Ri}|V_{Ri}|/2D) = \circ, \qquad (4.2.29)$$

$$c^+: (V_p - V_{Le}) + (\tfrac{g}{c})(H_p - H_{Le}) + (f.\Delta t)(f.V_{Le}|V_{Le}|/2D) = \circ, \qquad (4.2.30)$$

$$c^-: (V_p - V_{Ri}) + (\tfrac{g}{c})(H_p - H_{Ri}) + (f.\Delta t)(fV_{Ri}|V_{Ri}|/2D) = \circ, \qquad (4.2.31)$$

$$V_p = \frac{1}{2}\left[(V_{Le} + V_{Ri}) + \frac{g}{c}\left(H_{Le} - H_{Ri}\right) - (f.\Delta t/2D)(V_{Le}|V_{Le}| - V_{Ri}|V_{Ri}|)\right], \qquad (4.2.32)$$

$$H_p = \frac{1}{2}\left[\frac{c}{g}(V_{Le} + V_{Ri}) + (H_{Le} - H_{Ri}) - \frac{c}{g}(f.\Delta t/2D)(V_{Le}|V_{Le}| - V_{Ri}|V_{Ri}|)\right], \qquad (4.2.33)$$

$V_{Le}'\ V_{Ri}'\ H_{Le}'\ H_{Ri}'\ f\ D$ are initial conditions parameters.

They are applied for solution at steady state condition. Water hammer equations calculation starts with pipe length " L " divided by " N " parts:

$$\Delta S = \frac{L}{N} \ \& \ \Delta t = \frac{\Delta s}{C}$$

Equations (4.2.28) and (4.2.29) are solved for the range P_2 through P_N, therefore H and V are found for internal points. Hence:

At P_1 there is only one characteristic Line (c^-)

At P_{N+1} there is only one characteristic Line (c^+)

For finding H and V at P_1 and P_{N+1} the boundary conditions are used.

The challenge of selecting a time step is made difficult in pipeline systems by two conflicting constraints. They are defined by dynamic model for water hammer, first of all for calculating many boundary conditions, such as obtaining the head and discharge at the junction of two or more pipes. It is necessary that the time step be common to all pipes. The second constraint arises from the nature of the "MOC". If the adjective terms in the governing equations are neglected (as is almost always justified), the "MOC" requires that ratio of the distance x to the time step t be equal to the wave speed in each pipe. In other words, the Courant number should ideally be equal to one and must not exceed one by stability reasons.

TABLE 1 Eulerian based computational dynamic model.

Pipe	Length (m)	Diameter (mm)	Hazen-Williams Friction Coef.	Velocity (m/s)
P2	60.7	1200	90	2.21
P3	311	1200	90	2.21

TABLE 1   *(Continued)*

Pipe	Length (m)	Diameter (mm)	Hazen-Williams Friction Coef.	Velocity (m/s)
P4	1	1200	90	2.65
P5	0.5	1200	91	2.65
P6	108.7	1200	67	2.65
P7	21.5	1200	90	2.21
P8	15	1200	86	2.21
P9	340.7	1200	90	2.21
P10	207	1200	90	2.21
P11	339	1200	90	2.21
P12	328.6	1200	90	2.21
P13	47	1200	90	2.21
P14	590	1200	90	2.21
P15	49	1200	90	2.21
P16	224	1200	90	2.21
P17	18.4	1200	90	2.21
P18	14.6	1200	90	2.21
P19	12	1200	90	2.21
P20	499	1200	90	2.21
P21	243.4	1200	90	2.21
P22	156	1200	90	2.21
P23	22	1200	90	2.21
P24	82	1200	90	2.21
P25	35.6	1200	90	2.21
P0	0.5	1200	90	2.65
P1	0.5	1200	90	2.65

TABLE 2   Computational results of Eulerian based model.

Type of Point	Maximum Volume	Type of Point	Maximum Volume	Type of Point	Maximum Volume	Type of Point	Maximum Volume
P0:J1	Vapor	P12:J13	Vapor	P16:J16	Vapor	P21:J22	Vapor
P0:J2	Vapor	P13:33.33%	Vapor	P16:J17	Air	P22:10.00%	Vapor
P1:J2	Vapor	P13:66.67%	Vapor	P17:J17	Air	P22:20.00%	Vapor
P1:J3	Vapor	P13:J13	Vapor	P17:J18	Vapor	P22:30.00%	Vapor
P10:15.38%	Vapor	P13:J14	Vapor	P18:J18	Vapor	P22:40.00%	Vapor
P10:23.08%	Vapor	P14:10.81%	Vapor	P18:J19	Vapor	P22:50.00%	Vapor
P10:30.77%	Vapor	P14:13.51%	Vapor	P19:J19	Vapor	P22:60.00%	Vapor
P10:38.46%	Vapor	P14:16.22%	Vapor	P19:J20	Air	P22:70.00%	Vapor
P10:46.15%	Vapor	P14:18.92%	Vapor	P2:25.00%	Vapor	P22:80.00%	Vapor
P10:53.85%	Vapor	P14:2.70%	Vapor	P2:50.00%	Vapor	P22:90.00%	Vapor
P10:61.54%	Vapor	P14:21.62%	Vapor	P2:75.00%	Vapor	P22:J22	Vapor
P10:69.23%	Vapor	P14:24.32%	Vapor	P2:J6	Vapor	P22:J23	Vapor
P10:7.69%	Vapor	P14:27.03%	Vapor	P2:J7	Vapor	P23:50.00%	Vapor
P10:76.92%	Vapor	P14:29.73%	Vapor	P20:12.90%	Vapor	P23:J23	Vapor
P10:84.62%	Vapor	P14:32.43%	Vapor	P20:16.13%	Vapor	P23:J24	Vapor
P10:92.31%	Vapor	P14:35.14%	Vapor	P20:19.35%	Vapor	P24:20.00%	Vapor
P10:J10	Vapor	P14:37.84%	Vapor	P20:22.58%	Vapor	P24:40.00%	Vapor
P10:J11	Vapor	P14:40.54%	Vapor	P20:25.81%	Vapor	P24:60.00%	Vapor
P11:14.29%	Vapor	P14:43.24%	Vapor	P20:29.03%	Vapor	P24:80.00%	Vapor
P11:19.05%	Vapor	P14:45.95%	Vapor	P20:3.23%	Vapor	P24:J24	Vapor
P11:23.81%	Vapor	P14:48.65%	Vapor	P20:32.26%	Vapor	P24:J28	Air
P11:28.57%	Vapor	P14:5.41%	Vapor	P20:35.48%	Vapor	P25:33.33%	Vapor
P11:33.33%	Vapor	P14:51.35%	Vapor	P20:38.71%	Vapor	P25:66.67%	Vapor
P11:38.10%	Vapor	P14:54.05%	Vapor	P20:41.94%	Vapor	P25:J28	Air
P11:4.76%	Vapor	P14:56.76%	Vapor	P20:45.16%	Vapor	P25:N1	Vapor

TABLE 2   *(Continued)*

Type of Point	Maximum Volume	Type of Point	Maximum Volume	Type of Point	Maximum Volume	Type of Point	Maximum Volume
P11:42.86%	Vapor	P14:59.46%	Vapor	P20:48.39%	Vapor	P3:10.00%	Vapor
P11:47.62%	Vapor	P14:62.16%	Vapor	P20:51.61%	Vapor	P3:15.00%	Vapor
P11:52.38%	Vapor	P14:64.86%	Vapor	P20:54.84%	Vapor	P3:20.00%	Vapor
P11:57.14%	Vapor	P14:67.57%	Vapor	P20:58.06%	Vapor	P3:25.00%	Vapor
P11:61.90%	Vapor	P14:70.27%	Vapor	P20:6.45%	Vapor	P3:30.00%	Vapor
P11:66.67%	Vapor	P14:72.97%	Vapor	P20:61.29%	Vapor	P3:35.00%	Vapor
P11:71.43%	Vapor	P14:75.68%	Vapor	P20:64.52%	Vapor	P3:40.00%	Vapor
P11:76.19%	Vapor	P14:78.38%	Vapor	P20:67.74%	Vapor	P3:45.00%	Vapor
P11:80.95%	Vapor	P14:8.11%	Vapor	P20:70.97%	Vapor	P3:5.00%	Vapor
P11:85.71%	Vapor	P14:81.08%	Vapor	P20:74.19%	Vapor	P3:50.00%	Vapor
P11:9.52%	Vapor	P14:83.78%	Vapor	P20:77.42%	Vapor	P3:55.00%	Vapor
P11:90.48%	Vapor	P14:86.49%	Vapor	P20:80.65%	Vapor	P3:60.00%	Vapor
P11:95.24%	Vapor	P14:89.19%	Vapor	P20:83.87%	Vapor	P3:65.00%	Vapor
P11:J11	Vapor	P14:91.89%	Vapor	P20:87.10%	Vapor	P3:70.00%	Vapor
P11:J12	Vapor	P14:94.59%	Vapor	P20:9.68%	Vapor	P3:75.00%	Vapor
P12:14.29%	Vapor	P14:97.30%	Vapor	P20:90.32%	Vapor	P3:80.00%	Vapor
P11:33.33%	Vapor	P14:51.35%	Vapor	P20:38.71%	Vapor	P25:66.67%	Vapor
P11:38.10%	Vapor	P14:54.05%	Vapor	P20:41.94%	Vapor	P25:J28	Air
P11:4.76%	Vapor	P14:56.76%	Vapor	P20:45.16%	Vapor	P25:N1	Vapor
P12:19.05%	Vapor	P14:J14	Vapor	P20:93.55%	Vapor	P3:85.00%	Vapor
P12:23.81%	Vapor	P14:J15	Air	P20:96.77%	Vapor	P3:90.00%	Vapor
P12:28.57%	Vapor	P15:33.33%	Vapor	P20:J20	Air	P3:95.00%	Vapor
P12:33.33%	Vapor	P15:66.67%	Vapor	P20:J21	Vapor	P3:J7	Vapor
P12:38.10%	Vapor	P15:J15	Air	P21:12.50%	Vapor	P3:J8	Vapor
P12:4.76%	Vapor	P15:J16	Vapor	P21:18.75%	Vapor	P4:J3	Vapor

TABLE 2   *(Continued)*

Type of Point	Maximum Volume	Type of Point	Maximum Volume	Type of Point	Maximum Volume	Type of Point	Maximum Volume
P12:42.86%	Vapor	P16:14.29%	Vapor	P21:25.00%	Vapor	P4:J4	Vapor
P12:47.62%	Vapor	P16:21.43%	Vapor	P21:31.25%	Vapor	P5:J26	Air
P12:52.38%	Vapor	P16:28.57%	Vapor	P21:37.50%	Vapor	P5:J4	Vapor
P12:57.14%	Vapor	P16:35.71%	Vapor	P21:43.75%	Vapor	P6:14.29%	Vapor
P12:61.90%	Vapor	P16:42.86%	Vapor	P21:50.00%	Vapor	P6:28.57%	Vapor
P12:66.67%	Vapor	P16:50.00%	Vapor	P21:56.25%	Vapor	P6:42.86%	Vapor
P12:71.43%	Vapor	P16:57.14%	Vapor	P21:6.25%	Vapor	P6:57.14%	Vapor
P12:76.19%	Vapor	P16:64.29%	Vapor	P21:62.50%	Vapor	P6:71.43%	Vapor
P12:80.95%	Vapor	P16:7.14%	Vapor	P21:68.75%	Vapor	P6:85.71%	Vapor
P12:85.71%	Vapor	P16:71.43%	Vapor	P21:75.00%	Vapor	P6:J26	Air
P12:9.52%	Vapor	P16:78.57%	Vapor	P21:81.25%	Vapor	P6:J27	Vapor
P12:90.48%	Vapor	P16:85.71%	Vapor	P21:87.50%	Vapor	P7:50.00%	Vapor
P12:95.24%	Vapor	P16:92.86%	Vapor	P21:93.75%	Vapor	P7:J27	Vapor
P12:J12	Vapor	-	-	P21:J21	Vapor	-	-

Faced with this challenge, this work tried for ways of relaxing the numerical constraints.

For the velocity of surge or pressure wave in an elastic case with free water bubble, the flowing equation would be valid:

$$C = \cfrac{1}{\left[\rho \left(\left(\dfrac{1}{E_W} \right) + \left(\dfrac{D}{E.t_W} \right) + \dfrac{n}{P} \right) \right]^{\frac{1}{2}}}, \qquad (4.2.34)$$

For the velocity of surge or pressure wave in elastic case (Table 1) with the high value of free water bubble (Table 2) the flowing equation would be valid:

$$C = \left(\frac{g.h}{n} \right)^{1/2}, \qquad (4.2.35)$$

Really stopping of a second layer of liquid exerts pressure on the following layers gradually caused high wave pressure. It acts directly at the valve extends to the rest of the pipeline against fluid flow speed C.

If the pressure at the beginning of the pipeline remains unchanged then after the shock of the initial section of the tube, it begins the reverse movement of the shock wave with the same velocity C and instantaneous changes.

Instantaneous changes in the rate of flow at the pipeline causes the surge wave. This phenomena occurs during water hammer are explained on the basis of compressibility of liquid drops.

4.4   EXPERIMENTS IN ORDER TO PRESENTATION OF WATER HAMMER PHENOMENON

This models used for laboratory; computational and field tests experiments in order to presentation of water hammer phenomenon at the water pipeline. Field tests and laboratory experiments were performed at 0:00 hours on 02/10/2007–02/05/2009.

Consistency for the observed values of maximal pressure the corresponding values were calculated according to Zhukousky's formula. In the final procedure it was compared the results of Eulerian based computational model (4.2.1) through (4.3.1) with the results of field test model (4.2.36) for transient flow.

TABLE 3   Field tests and computational results of Eulerian based model in the water pipeline.

Type of Point	Percent of air volume	Velocity of surge wave (m/s)	Head (m)
P10	15.38%	113.6	129
P11	14.29%	94.9	126.1
P12	14.29%	96.8	133.1
P13	33.33%	63.5	132.7
P14	10.81%	121	146.2
P15	33.33%	66.5	146
P16	14.29%	102.4	146.9
P2	25.00%	79.2	156.8
P20	12.90%	110.3	145.9
P21	12.50%	105	143.1
P22	10.00%	118.4	140.2
P23	50.00%	53	140.4
P24	20.00%	83	137.8

TABLE 3   *(Continued)*

Type of Point	Percent of air volume	Velocity of surge wave (m/s)	Head (m)
P25	33.33%	56.2	104.3
P3	10.00%	115.5	133.3
P6	14.29%	106	157.3
P7	50.00%	56	157.2

4.5   REGRESSION MODEL DUE TO FIELD TEST FOR SURGE WAVE VELOCITY

Curve estimation is the most appropriate when the relationship between the dependent variable and the independent variable is not necessarily linear.

Linear regression is used to model the value of a dependent scale variable based on its linear relationship to one or more predictors.

Non-linear regression is appropriate when the relationship between the dependent and independent variables is not intrinsically linear. Binary logistic regression is most useful in modeling of the event probability for a categorical response variable with two outcomes.

The auto-regression procedure is an extension of ordinary least-squares regression analysis specifically designed for time series.

One of the assumptions underlying ordinary least-squares regression is the absence of auto-correlation in the model residuals. Time series, however, often exhibit first-order auto-correlation of the residuals. In the presence of auto-correlated residuals, the linear regression procedure gives inaccurate estimates of how much of the series variability is accounted for by the chosen predictors. This can adversely affect the choice of predictors, and hence the validity of the model. The auto-regression procedure accounts for first-order auto-correlated residuals. It provides reliable estimates of both goodness of-fit measures and significant levels of chosen predictor variables.

One of the approaches of this work was the definition of a model by regression. It was defined based on the relationship between the dependent and independent data or variables for surge wave. It was showed in (Figure 5) and Equation (4.2.36). The variables are as follows: c—Velocity of surge wave $\left(m/_s\right)$ as a dependent variable with nomenclature "Y". The independent variable with nomenclature "X" such as: n—Percent of air volume $\left(m^3\right)$.

The curve estimation procedure allows quick estimating regression statistics, and producing related plots for different models. Hence the auto-regression procedure by regression software "SPSS 10.0.5" was selected for the curve estimation procedure in the present work. Therefore the regression model was built based on the field test data.

Regression software "SPSS" fitted the function curve and provided regression analysis. So, the regression model (4.2.36) was found in the final procedure. By this model, laboratory and field test results were compared by computational model

(4.2.35). The main practical aim of the present work was concentrated on the definition of a condition base maintenance (CM) method for all water transmission systems.

The data collection procedures were as follows:

At fast transients, down to 1 s, surge pressure and velocity of surge wave were
recorded. These data were used for curve estimation procedure. They were detected
on actual systems (field tests). Also, flow and pressure were computed by computational model (4.2.32) and (4.2.33). These data were compared by flow and pressure
data which were collected from actual systems (field tests). The model was calibrated
using one set of data, without changing parameter values. It was used to match a different set of results. Curve estimation procedure (Figures 6–9) for surge wave velocity
was formed by estimating regression statistics. Although related plots for the field test
model were produced.

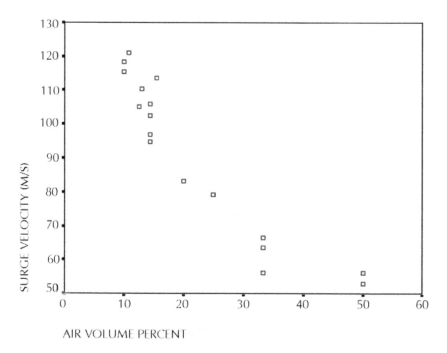

FIGURE 5   2-D Scatter diagram of surge wave speed for water pipeline with free water bubble.

TABLE 4   Regression for model Summary.

Model	R	R Square	Adjusted R Square	Std. Error of the Estimate
1	.930(a)	0.865	0.856	9.08222

[a]Predictors: (Constant), AIR VOLUME PERCENT

TABLE 5 Model summary for function of regression model.

Model		Sum of Squares	Df	Mean Square	F	Sig.
1	Regression	7934.64	1	7934.64	96.193	.000(a)
	Residual	1237.3	15	82.487		
	Total	9171.94	16			

a. Predictors: (Constant), AIR VOLUME PERCENT
b. Dependent Variable: SURGE VELOCITY(M/S)

TABLE 6 Variables and constant of function for Regression model.

Model		Unstandardized Coefficients		Standardized Coefficients	T	Sig.
		B	Std. Error	Beta		
1	(Constant)	127.451	4.35		29.301	0
Air volume percent	-1.673	0.171	-0.93	-9.808	0	

a. Dependent Variable: SURGE VELOCITY(M/S)

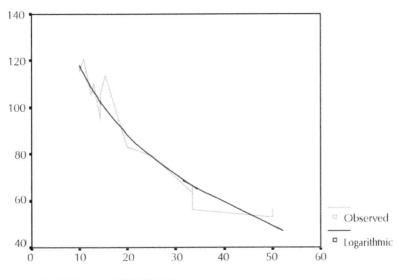

FIGURE 6 Logarithmic curve fit for surge wave speed of water pipeline with free water bubble.

TABLE 7 List wise deletion of missing data for curve fit by Logarithmic method.

Multiple R	.96799
R Square	. 93701
Adjusted R Square	.93281
Standard Error	6.20621

TABLE 8 Analysis of variance for curve fit by Logarithmic method.

	DF	Sum of Squares	Mean Square
Regression	1	8594.1825	8594.1825
Residuals	15	577.7563	38.5171

F = 223.12650 & Signif F = .0000

TABLE 9 Variables of the equation for curve fit by Logarithmic method.

Variable	B	SE B	Beta	T Sig T
VAR00002	−42.606130	2.852309	−.967992	−14.937
(Constant)	215.971673	8.522770		−25.341

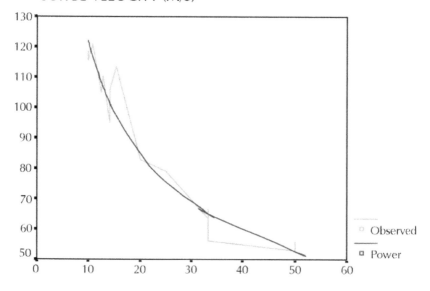

SURGE VELOCITY (M/S)

AIR VOLUME PERCENT

FIGURE 7 Power curve fit for surge wave speed of water pipeline with free water bubble.

TABLE 10   List wise deletion of missing data for curve fit by Power method.

Multiple R	.97433
R Square	.94932
Adjusted R Square	.94594
Standard Error	.06739

This work presented regression software "SPSS10.0.5" for fitting the function curve and providing regression analysis. So, by comparison of 10 methods (i.e. logarithmic method with Power method), finally the Power method (4.3.1) with standard error .06739 was selected as a regression model due to curve fitting (Figure 6).

TABLE 11   Analysis of variance for curve fit by Power method.

	DF	Sum of Squares	Mean Square
Regression	1	1.2761075	1.2761075
Residuals	15	.0681247	.0045416

F = 280.97904 & Signif F = .0000

TABLE 12   Curve fit by power method.

Variable	B	SE B	Beta	T Sig T
VAR00002	-.519175	.030973	-.974331	-16.762
(Constant)	402.197842	37.222066	-	10.805

TABLE 13   Variables of the equation for curve fit by Power method.

Dependent	Mth	Rsq	d.f.	F Sigf	b0	b1
VAR00001	POW	.949	15	280.98	402.198	-.5192

In power functions, however, a variable base is raised to a fixed exponent. The parameter $b_\circ$ serves as a simple scaling factor, moving the values of x^h up or down as $b_\circ$ increases or decreases, respectively the parameter b_1, called either the exponent or the power, determines the function's rates of growth or decay. Depending on whether it is positive or negative, a whole number or a fraction, b_1 will also determine the function's overall shape and behavior. So the flowing equation derived by the fitting function curve and regression analysis.

$$f(X) = b_\circ X^{b_1},\qquad\qquad (4.3.1)$$

$$f(X) = (402.198) * X^{(-.5192)}$$

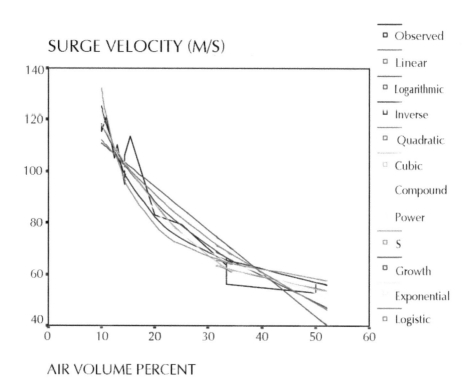

FIGURE 8 Comparison of curve fit for a series of functions due to surge wave in water pipeline.

TABLE 14 Field tests and computational results of Eulerian based model.

Type of Point	Percent of air volume (cm)	Type of Point	Percent of air volume (cm)
P10	92.31%	P20:J20	Air
P11	95.24%	P21	93.75%
P12	95.24%	P22	90.00%
P13	66.67%	P23	50.00%
P14	97.30%	P24	80.00%
P14:J15	Air	P24:J28	Air
P15	66.67%	P25	66.67%

TABLE 14   *(Continued)*

Type of Point	Percent of air volume (cm)	Type of Point	Percent of air volume (cm)
P15:J15	Air	P25:J28	Air
P16	92.86%	P3	95.00%
P16:J17	Air	P5:J26	Air
P17:J17	Air	P6	85.71%
P19:J20	Air	P6:J26	Air
P20	96.77%	P7	50.00%

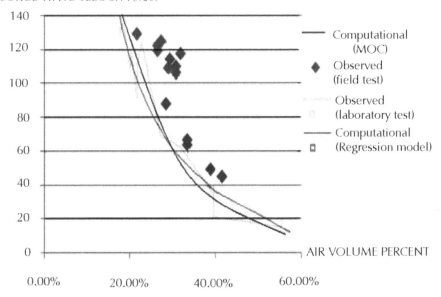

FIGURE 9   Comparison of computational results with laboratory and field test results.

This work selected the power function as a regression model (4.3.1) based on the field test metering. Hence it was compared with the computational results (4.2.35). It was assumed the penetrated air volume in the water pipeline which was showed in (Figure 9).In present work the equation that was offered by regression is as the flowing:

$$SurgeVelocity = 402.198 * (AirVolume)^{(-.5192)}$$

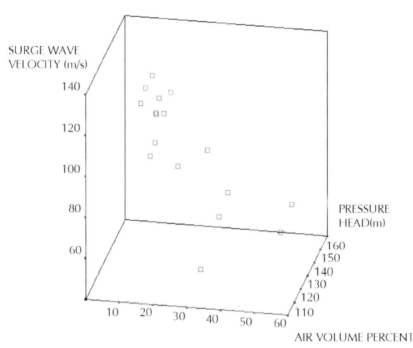

FIGURE 10   3-D Scatter diagram of surge wave speed for water pipeline with free water bubble.

Some of the reasons for changes at liquid properties in pipes and channels are due to the spread of the surge wave, the formation and collapse of vapor bubbles (cavitation) and air leak or disconnection of the system. The surge wave velocity has proportional changes against to changes in percent of the formation and collapse of vapor bubbles (Figure 9 and Figure 10).

The pipe system has a characteristic time period $T = \frac{2L}{a}$, where L is equal to 590 (m). It is the longest possible path through the system. The pressure wave speed a is equal to 1084 $\left(m/s \right)$. The time period is equal to 1.1 (s). It is equal to the time for a pressure wave to travel the pipe system's greatest length two times. Generally it is recommended that the run duration equals or exceeds T, which run duration or total simulation time was equal to 5(s). Run duration is measured either in seconds or as a number of time steps. Time steps typically range from a few hundredths of a second to a few seconds, depending on the system and the pressure wave speed. The run duration has a direct effect on the modeling computation time, along with the time step selected for the simulation. Another factor to consider definition of run duration is to allow enough time for friction to significantly dampen the transient energy at vapor pressure $-10.0\,(m)$.

4.5.1 Practical Implications

The conclusions were drawn on the basis of experiments and calculations for the pipeline with a local leak. Hence, the most important effects that were observed are as follows:

The pressure wave speed generated by water hammer phenomenon was influenced by some additional factors. Therefore the ratio of local leakage and discharge from the leak location was mentioned. The effect of total discharge from the pipeline and its effect on the values of wave oscillations period were studied. The outflow to the surge tank from the leak affected the value of wave celerity. The pipeline was equipped with the valve at the end of the main pipe, which was joined with the closure time register. The water hammer pressure characteristics were measured by extensometers, as similar as in the work of Kodura and Weinerowska. Simultaneously it was recorded in the computer's memory. The supply of the water at the system was realized with the use of the reservoir which enabled inlet pressure stabilization. In positive water hammer in the pipeline was introduced with local leaks in two scenarios; first; with the outflow

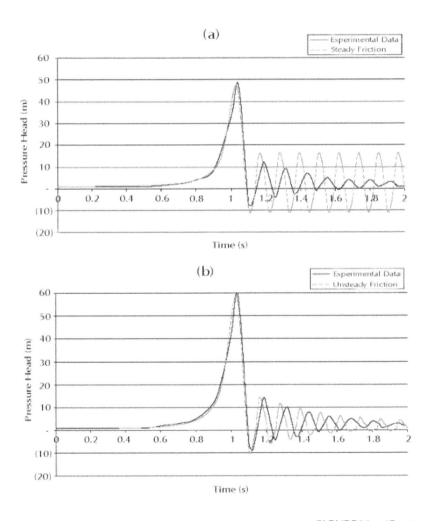

FIGURE11   *(Continued)*

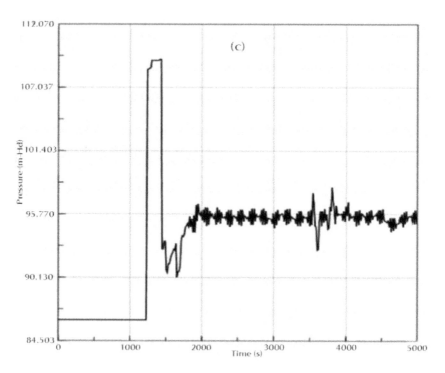

FIGURE 11 Pressure head histories for a single piping system using; (a) steady friction, (b) unsteady friction, Chaudhry, (c) steady friction, present computational results.

from the leak to the overpressure reservoir, second; with free outflow from the leak to atmospheric pressure, with the possibility of sucking air in the negative phase. The bubbles in non-linear dynamics fluid state acts as a separator gate for two distinct parts of the flow at upstream and downstream of separator gate. It causes high surge wave velocity at one of these distinct parts. Therefore the compressed air builds high pressure flow which can destroy the water pipeline.

Chaudhry [13], Obtained pressure heads by the steady and unsteady friction model (Figure 10). Comparison showed similarity in present work and work of Chaudhry [13], (Figures 11(a) and 11(b)) results. Parmakian [14], Streeter and Wylie [15].

The MOC based scheme was most popular because this scheme provided the desirable attributes of accuracy, numerical efficiency and programming simplicity in the work of Parmakian [14], Wylie and Streeter [15]. The water hammer software package was used in present chapter. The Numerical solution of the nonlinear Navier–Stokes equations and MOC was used by software package.

Leon [16] comparison showed similarity between present work results against the work of Leon [16].

The results are as following:
1. Numerical tests showed that the proposed second-order formulation at boundary conditions (achieved by using virtual cells) is second-order. In addition,

the proposed formulation maintains the conservation property of FV schemes and introduced no unphysical perturbations into the computational domain.

2. Numerical tests were performed for smooth (i.e., flows that do not present discontinuities) and sharp transients.

3. The high efficiency of the proposed scheme was important for real-time control (RTC) of water hammer flows in large networks.

Changes at system boundaries (Sudden changes) created a transient pressure pulse. In this regard, model design needed to find the relation between many variables accordance to fluid transient. Therefore, a computational technique was presented and the results were compared by field tests. In present work after closing the valves on the horizontal pipe of constant diameter, which moves the liquid with an average speed V_o, a liquid layer, located directly at the gate, immediately stops. Then successively terminate movement of the liquid layers (turbulence, counter flows) to increase with time away from the gate. In this work the air was sucked into the pipeline. Pressure wave velocity was recorded in fast transient up to 5 ms (in this work 1 second). The assessment procedure was used to analysis the collected data which were obtained at real system by Hariri et al. [17].

4.6 NAVIER-STOKE'S MODELING FOR NARROW PIPES

Water hammer numerical modeling and simulation processes were presented by three models:

1. Laboratory Model,
2. MOC Model,
3. Regression Model.

Three dimensional heat transfer characteristics and pressure drop of water flow in a set of rectangular mictrotubes were numerically investigated. A FV solver was employed to predict the temperature field for the conjugate heat transfer problem in both the solid and liquid regions in the microchannels. The full Navier–Stoke's approach was examined for this kind of narrow tubes for the pressure drop evaluations. The complete form of the energy equation with the dissipation terms was also linked to the momentum equations. The computed thermal characteristics and pressure drop showed good agreements with the experimental data. The effects of flow rate and channel geometry on the heat transfer capability and pressure drop of the system were also predicted.

To design an effective microtubes heat sink, fundamental understanding of the characteristics of heat transfer and fluid flow in microchannels are necessary. At the early stages, the designs and relations of macroscale fluid flow and heat transfer were employed. However, many experimental observations in microchannels deviate significantly from those in macroscale channels.

These disagreements were first observed by Tuckerman and Pease [18]. They demonstrated that micro rectangular passages have a higher heat transfer coefficient in laminar regimes in comparison with turbulent flow through macroscale channels. Therefore, they are capable of dissipating significant high heat fluxes.

Since, the regime of the flow has a noticeable influence on studying determination of heat dissipation rate; several researches have been conducted in this field. Wu and Little [19] measured the heat transfer characteristics for gas flows in miniature channels with inner diameter ranging from 134 (μm) to 164 (μm). The tests involved both laminar and turbulent flow regimes. Their results showed that the turbulent convection occurs at Reynolds number of approximately 1,000.

They also found that the convective heat transfer characteristics depart from the predictions of the established empirical correlations for the macroscale tubes. They attributed these deviations to the large asymmetric relative roughness of the micro-channel walls. Harms et al. [20] tested a 2.5 (cm) long, 2.5 (cm) wide silicon heat sink having 251(μm) wide and 1030 (μm) deep microchannels. A relatively low Reynolds number of 1,500 marked transition from laminar to turbulent flow which was attributed to a sharp inlet, relatively long entrance region, and channel surface roughness. They concluded the classical relation for Nusselt number was fairly accurate for modeling microchannel flows.

Fedrov and Viskanta [21] reported that the thermal resistance decreases with Reynolds number and reaches an asymptote at high Reynolds numbers.

Qu et al. [22] investigated heat transfer and flow characteristics in trapezoidal silicon microtubes. In comparison, the measured friction factors were found to be higher than the numerical predictions. The difference was attributed to the wall roughness. Based on a roughness-viscosity model, they explained that the numerically predicted Nusselt numbers are smaller than the experimentally determined ones.

Choi et al. [23] measured the convective heat transfer coefficients for flow of nitrogen gas in microtubes for both laminar and turbulent regimes. They found that the measured Nusselt number in laminar flow exhibits a Reynolds number dependence in contrast with the conventional prediction for the fully developed laminar flow, in which Nusselt number is constant.

Adam et al. [24] conducted single-phase flow studies in microtubes using water as the working fluid. Two diameters of the circular microtubes, namely 0.76 (mm) and 1.09 (mm), were used in the investigation. It was found that the Nusselt numbers are larger than those encountered in microtubes. Peng and Peterson [25] investigated water flows in rectangular microtubes with hydraulic diameters ranging from 0.133 to 0.336 (mm). In laminar flows, it was found that the heat transfer depends on the aspect ratio and the ratio of the hydraulic diameter to the center-to-center distance of the microchannel.

Mala et al. [26] considered the electrical body forces resulting from the double layer field in the equations of motion. These effects are negligible in the macroscale as the dimensions of the electric double layer, EDL, is very small with respect to tube dimensions. They solved the PoissonBolzmann equation for the steady state flow. It was found that without the double layer a higher heat transfer rate is obtained. They proposed to consider the effects of the EDL on liquid flows and heat transfer in micro-tubes to prevent the overestimation of the heat transfer capacity of the system. Xu et al. [27] investigated the effects of viscous dissipation on the micro scale dimensions.

They used a 2D microtube and considered the viscous dissipation term in energy equation. The results show that this term plays a significant role in temperature, pressure and velocity distributions. Therefore, the relationships between the friction factor and the Reynolds number change when the hydraulic diameter of the micrrotube is very small. The viscous dissipation effects are brought about by rises in the velocity gradient as hydraulic diameter reduces for a constant Reynolds number, a numerical study is conducted based on the experimental results of Tuckerman [28]. A FV method is used to solve the conjugate heat transfer through the heat sinks. The flow and heat transfer development regions inside the tubes are considered. The numerical results are then compared with the available experimental data. The effects of liquid velocity through channels and their effects on heat transfer and pressure drop along microchannels are investigated. Finally, the effects of aspect ratio on heat dissipation and pressure drop in microtubes are predicted.

The characteristics of the model are depicted in (Figure 12). The width and thickness of microchannels are W_1, and H_1 respectively. The thickness of the silicon substrate is H_2 -H_1; and the total length of the microchannels is L. The heat supplies by a 1 × 1 cm heat source located at the entrance of the channels and were centered across the whole channel heat sink. A uniform heat flux of q is provided to heat the microchannels. The heat is removed by flowing water through channels. The inlet temperature of the cooling water is 20°C. The analysis is performed for five different cases.

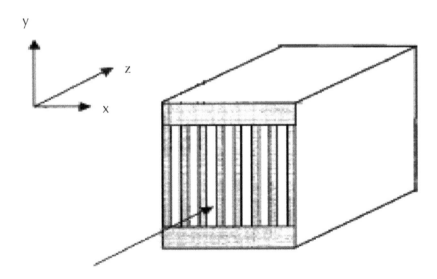

FIGURE 12 A sample microtube.

The dimensions related to each case are given in Table 15. By these dimensions; there will be 150 microchannels for cases 0 and 1 and 200 microchannels for cases 2, 3, and 4.

For the second section, case 2 was considered as the base geometry. The water velocity changed from 50 (cm/s) to 400 (cm/s).

In the third section, the Tuckerman's geometries were solved by the unique Reynolds of 150. As performing a numerical method for the whole microtubes heat sink is hard; a certain computational domain is considered (Figure 13).

To prevent the various boundary condition effects, the computational domain is taken at the center of the heat sink to have the quoted uniform heat flux in Table 15. This is because there is very little spreading of heat towards the heat sinks. There is also some geometrical symmetry which simplifies the computation. So only a semi-channel and semi-silicon substrate will be considered and the results will be the same for the other half. The whole substrate is made of silicon with thermal conductivity (k) of 148 (W/m. K). At the top of the channel $y = H_2$, there is a Pyrex plate to make an adiabatic condition (its thermal conductivity is two orders lower than silicon). There are two different boundary conditions at the bottom.

TABLE 15   Four different cases of micro channels.

	Case				
	0	*1*	*2*	*3*	*4*
L(cm)	2	2	1.4	1.4	1.4
$W_1(\mu m)$	64	64	56	55	50
$W_2(\mu m)$	36	36	44	45	50
$H_1(\mu m)$	280	280	320	287	302
$H_2(\mu m)$	489	489	533	430	458
$Q(cm^3/s)$	1.277	1.86	4.7	6.5	8.6
$q(W/cm^2)$	34.6	34.6	181	277	790
Number of channels	150	150	200	200	200

For $z < L_h$ a uniform heat flux of q is imposed over the heat sink and the rest is assumed to be adiabatic. Water flows through the channel from the entrance in z direction.

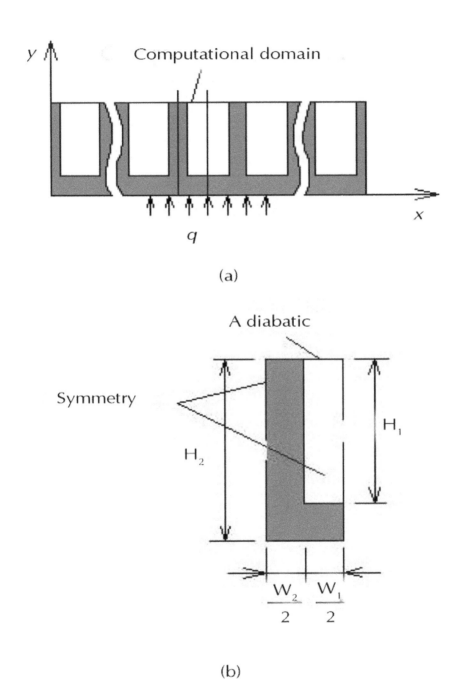

FIGURE 13 Dimensions and computational domain.

4.6.1 Boundary Conditions

The transverse velocities of the inlet are assumed to be zero. The axial velocity is considered to be evenly distributed through the whole channel. The velocities at the top and bottom of tubes are zero.

4.6.2 Governing Equations

Several simplifying assumptions are incorporated before establishing the governing equations for the fluid flow and heat transfer in a unit cell:

1. Steady fluid flow and heat transfer;
2. Incompressible fluid;
3. Laminar flow;
4. Negligible radiative heat transfer;
5. Constant solid and fluid properties.

In the last assumption the solid and liquid properties are assumed to be constant because of the small variations within the temperature range tested. .

Based on the above assumptions, the governing differential equations and Nomenclatures used to describe the fluid flow and heat transfer in a unit cell are expressed as follows:

A = Flow area, T = Temperature, cp = Specific heat, Tin = Inlet temperature, Dh = Hydraulic diameter, T_{max} = Maximum temperature, EDL = Electric Double Layer, w = Velocity (z direction), H_1 = Channel height, W_1 = Channel width, H_2 = Total height, W_2 = Substrate width, k = Thermal conductivity, L = Heat sink length, Φ = Viscous dissipation terms in energy equation, Lh = Heated length, p = Pressure,μ = Viscosity, P = Wetted perimeter, ρ = Density, q = Heat flux subscripts, Q = Total average volumetric Flow rate, R = Thermal resistance ,in = Inlet, Re = Reynolds number, max = Maximum.

Continuity Equation

$$\frac{\partial w}{\partial z} = 0 \qquad (4.5.1)$$

Momentum Equation

$$\rho \frac{\partial (w_i w_j)}{\partial z_j} = -\frac{\partial p}{\partial z_i} + \mu \frac{\partial}{\partial z_j}\left(\frac{\partial w_i}{\partial z_j} + \frac{\partial w_j}{\partial z_i}\right)$$

$$-\frac{2}{3}\mu \frac{\partial}{\partial z_i}\left(\frac{\partial w_k}{\partial z_k}\right) \qquad (4.5.2)$$

Energy Equation

$$\rho c_p \frac{\partial w_j T}{\partial z_j} = k \frac{\partial^2 T}{\partial x_j^{\,2}} + \mu \Phi \qquad (4.5.3)$$

where

$$\Phi = 2\left[\left(\frac{\partial u}{\partial x}\right)^2 + \left(\frac{\partial v}{\partial y}\right)^2 + \left(\frac{\partial w}{\partial z}\right)^2\right] + \left(\frac{\partial u}{\partial y} + \frac{\partial v}{\partial x}\right)^2$$

$$+ \left(\frac{\partial u}{\partial z} + \frac{\partial w}{\partial x}\right)^2 + \left(\frac{\partial v}{\partial z} + \frac{\partial w}{\partial y}\right)^2 \tag{4.5.4}$$

4.6.3 Numerical Method

The finite volume method (FVM) is used to solve the continuity, momentum, and energy equations. A very brief description of the method used is given here.

In this method the domain is divided into a number of control volumes such that there is one control volume surrounding each grid point. The grid point is located in the center of a control volume. The governing equations are integrated over the individual control volumes to construct algebraic equations for the discrete dependent variables such as velocities, pressure, and temperature. The discretization equation then expresses the conservation principle for a finite control volume just as the partial differential equation expresses it for an infinitesimal control volume. In the present study, a solution is deemed converged when the mass imbalance in the continuity equation is less than 10^{-6}.

Section 1, Comparison of the Numerical Method with the Experimental Data

As the thermal specifications and flow characteristics are of the great importance in design of microchannel heat sinks, the results are concentrated in these fields.

For solving the equations several grid structures were used. The grid density of $120 \times 40 \times 20$ in z, y, and x directions is considered to be appropriate.

The thermal resistance is calculated as follows:

$$R(z) = \frac{T_{max}(z) - T_{in}}{q} \tag{4.5.5}$$

In Equation (4.5.5.), $R(z), T_{max}(z)$, T_{in} and q the thermal resistance at z (cm) from the entrance, the inlet water temperature and the heat flux at the heating area.

In addition to thermal resistance, the Reynolds number is calculated as:

$$Re \equiv \frac{\rho w_{ave} D_h}{\mu} \tag{4.5.6}$$

where

$$D_h \equiv \frac{4A}{P} = \frac{2H_1 W_1}{H_1 + W_1} \tag{4.5.7}$$

First, the temperature distribution, thermal resistances, and pressure drops based on the numerical results are demonstrated. Then the numerical results are compared with Tuckerman's experiments data.

Temperature Distribution and Thermal Resistance

The temperature distributions at four x-y cross sections along the channel are shown in Figure 14 for Case 0. The four sections, z_ 1, 3, 6, 9(mm) are all in the heated area. As all heat exchangers, isotherms are closest at the entrance of the channels. This means that heat transfer rate is the highest at the entrance and it decreases along the channel.

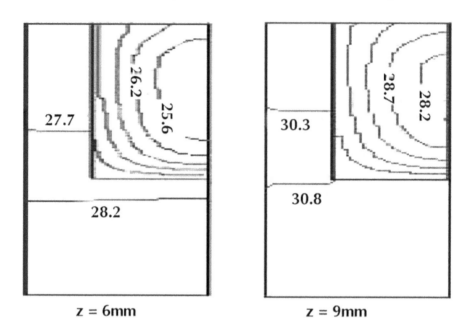

z = 6mm   z = 9mm

FIGURE 14   Temperature distribution along the tube for case 0.

The thermal resistance along the tube is shown in Figure 14 for cases 0 and 1. It can be seen that the thermal resistance has increased by increasing the z value and attained the maximum value at z = 9(mm). It is consistent with Tuckerman's experiments. It may be seen that the thermal resistance increased linearly through the channel except the entry region and near its peak. The sharper slopes in the experimental data of Figure 15(a) and the numerical prediction in Figure 15(b) are evidently due to the entrance region effects. So the flow may be considered thermally fully developed with a proper precision. The maximum thermal resistance is occurred in z = 9(mm) which is consistent with Tuckerman's experiments. For the other Cases the results are given in Table 16. The numerical values of resistances are predicted well.

Pressure Drop

Pressure drop is linear along the channel. Figure 16(a) shows the pressure drop for Case 4. The pressure drop increases by increasing the inlet velocity. The slope of the pressure line in the entrance of the channel is maximum. This is due to the entry region effects. The velocity field will be fully developed after a small distance from the entrance. So the assumption of fully developed flow is acceptable. Figure 16(b) shows the pressure drop for all 5 cases. These amounts are tabulated in Table 4.5.3.

TABLE 16   Thermal resistance comparison.

Case	$q(\dfrac{W}{cm^2})$	$R(cm^2 K/W)$		Error (%)
		Experimental	Numerical	
0	34.6	0.277	0.253	5.8
1	34.6	0.280	0.246	12.1
2	181	0.110	0.116	5.0
3	277	0.113	0.101	8.1
4	790	0.090	0.086	3.94

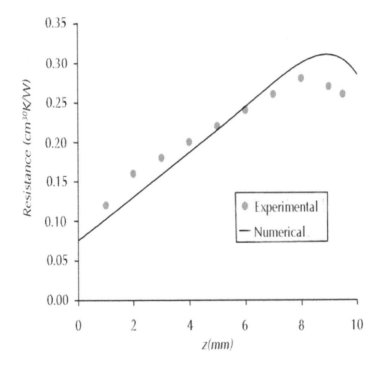

(a)

FIGURE 15   *(Continued)*

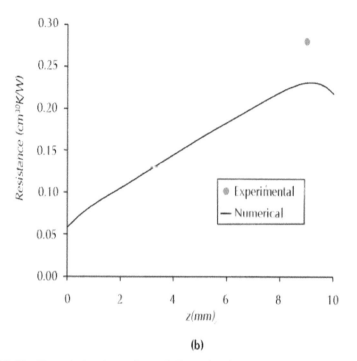

(b)

FIGURE 15   Numerical and experimental Thermal resistances for: (a) Case 0, and (b) Case 1.

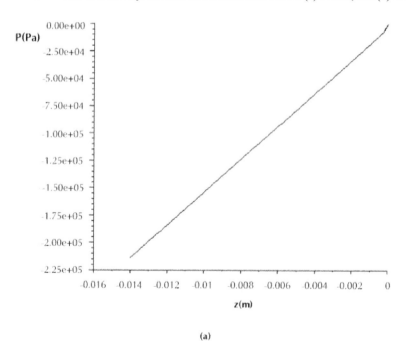

(a)

FIGURE 16   *(Continued)*

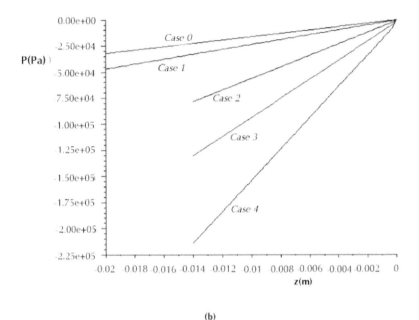

(b)

FIGURE 16   Pressure drop, (a) case 4, (b) all cases.

TABLE 17   Pressure drop in 5 Cases.

Case	Pressure Drop (bar)
0	0.322
1	0.469
2	0.784
3	1.302
4	2.137

Section 2, the Effect of Velocity on Temperature Distribution and Pressure Drop
The effects of velocity on the temperature rise and pressure drop in microtubes are displayed in Figure 17 and Figure 18. The amount of heat dissipation increases by increasing the velocity. But it can be seen that the amount of decrease in temperature falls drastically by increase in velocity.

The following function can predict the temperature rise:

$$(T_{max} - T_{in}) = 265.67 \ w^{-0.4997} \tag{4.5.8}$$

The pressure drop is a linear function of velocity. The amounts of temperature rise and pressure drop are given in Table 18.

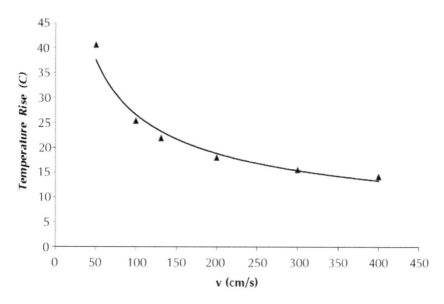

FIGURE 17   Temperature rise for different velocities.

TABLE 18   Temperature rise and pressure drop for section 2.

Velocity (cm/s)	Re	Temp. Rise (°C)	Pressure Drop (bar)
50	47	40.6	0.298
100	95	25.4	0.598
131	124	21.9	0.784
200	190	18.0	1.206
300	285	15.5	1.817
400	380	14.2	2.439

Section 3, the Effect of Aspect Ratio on Temperature Distribution and Pressure Drop

Aspect ratio is an important factor in microtube design. In this section the inlet heat flux of 181 (W/cm^2) imposed over the heat sinks and the hydraulic diameter changes between 85.8 and 104.2. (Figure 19) shows the maximum temperature of each case. It may be seen that with an identical heat flux the heat dissipation of the largest aspect ratio is the lowest. But this case has the minimum pressure drop too (Figure 20). By increasing the amount of aspect ratio the ability of heat dissipation increases.

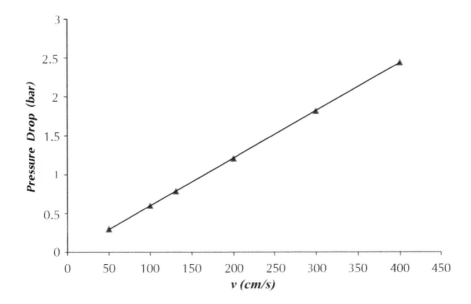

FIGURE 18   Pressure drop for different velocities.

TABLE 19   Temperature rise and pressure drop for section 3.

Case	Aspect Ratio	Hydraulic Diameter (µm)	Velocity (m/s)	Temp. Rise (°C)	Pressure Drop (bar)
1	4.375	104.19	1.447	27.84	0.98
2	5.714	95.32	1.58	19.97	0.95
3	5.218	92.31	1.633	20.76	1.03
4	6.040	85.80	1.76	20.35	1.31

The third geometry has the minimum temperature rise. The temperature rise of second and forth geometries are approximately alike. But the pressure drop of these two cases has a noticeable difference. The pressure drop increases with increase in aspect ratio. The amounts of temperature rise and temperature rise of all geometries are given in Table 19.

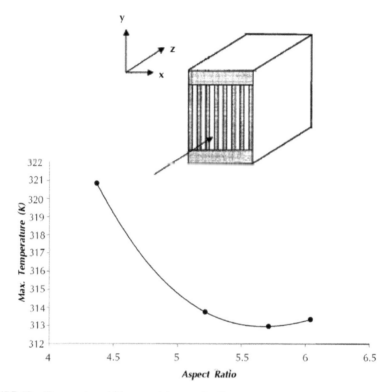

FIGURE 19   Temperature with respect to aspect ratio.

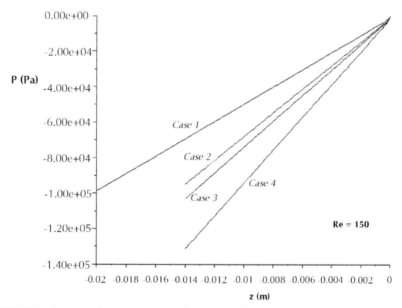

FIGURE 20   Pressure drop relating to each geometry in constant Reynolds of 150.

4.7 CONCLUSION

Generally, water hammer is manifested as a hydro-machines phenomenon which can leading to the destruction of pipelines. The cycles of increased and decreased in pressure iterates at intervals equal to time for dual-path shock wave length of the pipeline from the valve prior to the pipeline. Thus, the hydraulic impact of the liquid in the pipeline performed oscillatory motion. The cause of oscillatory motion was the hydraulic resistance and viscosity. It absorbed the initial energy of the liquid for overcoming the friction.

The effects of the penetrated air on the surge wave velocity in water pipeline. It showed that Eulerian based computational model is more accurate than the regression model. Hence in order to presentation for importance of penetrated air on water hammer phenomenon, it was compared the models for laboratory; computational and field tests experiments. As long as these procedure, it was showed that the Eulerian based model for water transmission line in comparison with the regression model. On the other hand, this idea were included the proper analysis to provide a dynamic response to the shortcomings of the system. It also performed the design protection equipments to manage the transition energy and determine the operational procedures to avoid transients. Consequently, the results will help to reduce the risk of system damage or failure at the water pipeline.

KEYWORDS

- **Eulerian based model**
- **Method of characteristics**
- **Navier–Stokes equations**
- **Surge wave**
- **Water transmission**

REFERENCES

1. Joukowski, N. Paper to Polytechnic Soc. Moscow, spring of 1898. *English translation by Miss O. Simin. Proc. AWWA* 57–58 (1904).
2. Hariri Asli, K., Nagiyev, F. B., and Haghi, A. K. A computational method to Study transient flow in binary mixtures. *Computational Methods in Applied Science and Engineering.* Chapter 13, Nova Science Publications ISBN: 978-1-60876-052-7, USA, pp. 229–236, https://www.novapublishers.com/catalog/ (2010).
3. Parmakian, J. Water hammer Design Criteria. *J. of Power Div., ASCE,* 456–460 (September, 1957).
4. Wood, D. J., Dorsch, R. G., and Lightener, C. Wave-Plan Analysis of Unsteady Flow in Closed Conduits. *Journal of Hydraulics Division, ASCE,* **92** 83–110 (1966).
5. Wylie, E. B. and Streeter, V. L. *Fluid Transients in Systems.* Prentice-Hall, Englewood Cliffs, New Jersey, p. 4 (1993).
6. Brunone, B., Karney, B. W., Mecarelli, M., and Ferrante, M. Velocity Profiles and Unsteady Pipe Friction in Transient Flow. *Journal of Water Resources Planning and Management, ASCE,* **126**(4) 236–244 (July, 2000).

7. Koelle, E., Luvizotto, Jr. E., and Andrade, J. P. G. *Personality Investigation of Hydraulic Networks using MOC – Method of Characteristics*. Proceedings of the 7th International Conference on Pressure Surges and Fluid Transients, Harrogate Durham, United Kingdom, pp. 1–8 (1996).

8. Filion, Y. and Karney, B. W. *A Numerical Exploration of Transient Decay Mechanisms in Water Distribution Systems*. Proceedings of the ASCE Environmental Water Resources Institute Conference, American Society of Civil Engineers, Roanoke, Virginia, p. 30 (2002).

9. Hamam, M. A. and Mc Corquodale, J. A. Transient Conditions in the Transition from Gravity to Surcharged Sewer Flow. *Canadian J. of Civil Eng., Canada* 65–98 (September, 1982).

10. Savic, D. A. and Walters G. A. *Genetic Algorithms Techniques for Calibrating Network Models*. Report No. 95/12, Centre for Systems and Control Engineering, pp. 137–146 (1995).

11. Walski, T. M., and Lutes, T. L. Hydraulic Transients Cause Low-Pressure Problems. *Journal of the American Water Works Association*, **75**(2) 58 (1994).

12. Lee, T. S. and Pejovic, S. Air influence on similarity of hydraulic transients and vibrations. *ASME J. Fluid Eng.*, **118**(4) 706–709 (1996).

13. Chaudhry, M. H. *Applied Hydraulic Transients*. Van Nostrand Reinhold Co., New York, pp. 1322–1324 (1979).

14. Parmakian, J. *Water hammer Analysis*. Dover Publications, Inc., New York, pp. 51–58 (1963).

15. Streeter, V. L. and Wylie, E. B. *Fluid Mechanics*. McGraw-Hill Ltd., USA, pp. 492–505 (1979).

16. Leon Arturo, S. An efficient second-order accurate shock-capturing scheme for modeling one and two-phase water hammer flows. PhD Thesis, pp. 4–44 (March 29, 2007).

17. Hariri Asli, K., Nagiyev, F. B., and Haghi, A. K. Some Aspects of Physical and Numerical Modeling of water hammer in pipelines. *Nonlinear Dynamics an International Journal of Nonlinear Dynamics and Chaos in Engineering Systems*, ISSN: 1573-269X (electronic version) Journal no. 11071 Springer, Netherlands, ISSN: 0924-090X (print version), Springer, Heidelberg, Germany, Number 4, Vol. 60, pp. 677–701, http://www.springerlink.com/openurl.aspgenre=article&id=doi:10.1007/s11071-009-9624-7 (2009, June, 2010).

18. Tuckerman, D. B. and Pease, R.F.W. High performance heat sinking for VLSI. *IEEE Electron device letter, DEL*, **2** 126–129 (1981).

19. Wu, P. Y. and Little, W. A. Measurement of friction factor for flow of gases in very fine channels used for micro miniature. *Joule Thompson refrigerators, Cryogenics*, **24**(8) 273–277 (1983).

20. Harms, T. M., Kazmierczak, M. J., Cerner, F. M., Holke, A., Henderson, H. T., Pilchowski, H. T., and Baker, K. Experimental Investigation of Heat Transfer and Pressure Drop through Deep Micro channels in a (100) Silicon Substrate, in: Proceedings of the ASME. *Heat Transfer Division, HTD*, **351** 347–357 (1997).

21. Fedorov, A. G. and Viskanta, R. Three-dimensional Conjugate Heat Transfer into Microchannel Heat Sink for Electronic Packaging. *Int. J. Heat Mass Transfer*, **43** 399–415 (2000).

22. Qu, W., Mala, G. M., and Li, D. *Heat Transfer for Water Flow in Trapezoidal Silicon Microchannels*, pp. 399–404 (1993).

23. Choi, S. B., Barren, R. R., and Warrington, R. O. Fluid Flow and Heat Transfer in Microtubes. *ASME DSC*, **40** 89–93 (1991).

24. Adams, T. M., Abdel-Khalik, S. I., Jeter, S. M., and Qureshi, Z. H. An Experimental investigation of single-phase Forced Convection in Microchannels. *Int. J. Heat Mass Transfer*, **41** 851–857 (1998).

25. Peng, X. F. and Peterson, G. P. Convective Heat Transfer and Flow Friction for Water Flow in Microchannel Structure. *Int. J. Heat Mass Transfer*, **36** 2599–2608 (1996).
26. Mala, G., Li, D., and Dale, J. D. Heat Transfer and Fluid Flow in Microchannels. *J. Heat Transfer*, **40** 3079–3088 (1997).
27. Xu, B., Ooi, K. T., Mavriplis, C., and Zaghloul, M. E. *Viscous dissipation effects for liquid flow in microchannels, Microsystems* pp. 53–57 (2002).
28. Tuckerman, D. B. and Pease, R. F. W. High performance heat sinking for VLSI. *IEEE Electron device letter, DEL*, **2** 126–129 (1981).

5 Computer Models for Fluid Interpenetration

CONTENTS

5.1 INTRODUCTION

Present book discusses on Eulerian based model for water hammer. This model was defined by the method of characteristics "MOC", finite difference form. The method was encoded into an existing hydraulic simulation model. The surge wave was assumed as a failure factor in an elastic case of water pipeline with free water bubble. The results were compared by regression analysis. It indicated that the accuracy of the Eulerian based model for water transmission line.

Miscible fluid mixtures that formed in the interpenetration between water flows at the process of interpenetration of two fluids at parallel between plates and turbulent moving in pipe are common phenomena in our daily life. While two miscible fluids are mixed together, their appearances in terms of colors and shapes will change due to their mixing interpenetration. The interpenetration between the mixture components could be regarded as a combination of the diffusing process and remixing process. If the former dominates the interpenetration, it is miscible, otherwise, it is immiscible.

5.2 COMPUTER MODELS FOR FLUID INTERPENETRATION

Water columns typically separate at abrupt changes in profile or local high points due to sub atmospheric pressure. The space between the water columns is filled either by the formation of vapor (e.g., steam at ambient temperature) or air, if it is admitted to the pipeline through a valve. The complex microscopic interplay between the mixture components makes the simulation highly challenging. So far, there have been some dedicated research in computer graphics dealing with immiscible mixtures, but few works have been done focusing on miscible mixtures.

Changes in fluid properties such as depressurization due to the sudden opening of a relief valve, a propagating pressure pulse, heating or cooling in cogeneration or industrial systems, mixing with solids or other liquids (may affect fluid density, specific gravity, and viscosity), formation and collapse of vapor bubbles (cavitations), and air entrainment or release from the system (at air vents and/or due to pressure waves). Changes at system boundaries such as rapidly opening or closing a valve, pipe burst (due to high pressure), or pipe collapse (due to low pressure), pump start/shift/stop, air intake at a vacuum breaker, water intake at a valve, mass outflow at a pressure-relief valve or fire hose, breakage of a rupture disk, and hunting and/or resonance at a control valve. Sudden changes such as these create a transient pressure pulse that rapidly propagates away from the disturbance in every possible direction and throughout the entire pressurized system. If no other transient event is triggered by the pressure wave fronts, unsteady-flow conditions continue until the transient energy is completely damped and dissipated by fiction.

The majority of transients, in water and wastewater systems are the result of changes at system boundaries, typically at the upstream and downstream ends of the system or at local high points. Consequently, results of present research can reduce the risk of system damage or failure with proper analysis to determine the system's default dynamic response, design protection equipment to control transient energy, and specify operational procedures to avoid transients. Analysis, design, and operational procedures all benefit from computer simulations in this research.

The study of hydraulic transients began with the work of Zhukovsky [1] and Allievi [2]. Many researchers have made significant contributions in this area, including Wood [3], Angus and Parmakian [4, 5], who popularized and perfected the graphical method of calculation. Benjamin Wylie and Victor Streeter [6-11], method of characteristics combined with computer modeling. The subject of transients in liquids are still growing fast around the world. Brunone et al [12], Koelle and Luvizotto [13], Filion and Karney [14], Hamam and McCorquodale [15], Savic and Walters [16, 17], Walski and Lutes [18], Wu and Simpson [19], have been developed various methods of investigation of transient pipe flow. These ranges of methods are included by approximate equations to numerical solutions of the nonlinear Navier–Stokes equations. Several approaches have been taken to numerically model the movement and transformation of pressure waves in water transmission systems and can be classified as Eulerian or Lagrangian. The Eulerian approach reformulates the governing transient flow equations, which are taken, expressed in a "MOC", finite difference form and regression method. The autoregression procedure accounts for first-order autocorrelated residuals and provides reliable estimates of both goodness-of-fit measures and significance levels of chosen predictor variables.

For this work data were recorded by "PLC". Research focused on reducing of unaccounted for water "UFW" where possible to reduce energy costs and so it has ensured reliable water transmission for the Rasht city main pipeline. It referred to a fluid transient as a "Dynamic" operating case, which may also include sudden thrust due to relief valves that pop open or rapid piping accelerations due to an earthquake. It is advisable to investigate fluid-structure interpenetration (FSI). Model design need to find the relation between two or many of variables accordance to fluid transient as a "Dynamic" operating.

Water hammer numerical modeling and simulation processes were handled by two models:

(1 First Model Approached to Transient Flow (Regression Model).

(2) Second Model Approached to Transient Flow (Method of characteristics "MOC" Model).

The regression software fitted the function curve and provided regression analysis. So in the final procedure it qualified a regression model. This model was compared with "MOC" model. This was the main practical aim of present PhD research. So, this model became a Condition Base Maintenance (CBM) manifest. The curve estimation procedure allowed quickly estimating regression statistics and producing related plots for different models. Curve estimation was most appropriate when the relationship between the dependent variable(s) and the independent variable(s) was not necessarily linear. Linear regression was used to model the value of a dependent scale variable based on its linear relationship to one or more predictors. Nonlinear regression was appropriate when the relationship between the dependent and independent variables was not intrinsically linear. Binary logistic regression was most useful in modeling of the event probability for a categorical response variable with two outcomes. The auto regression procedure was an extension of ordinary least squares regression analysis specifically designed for time series. One of the assumptions underlying ordinary least squares regression was the absence of autocorrelation in the model residuals. Time series, however, often exhibit first-order autocorrelation of the residuals. In the presence of auto correlated residuals, the linear regression procedure gave inaccurate estimates of how much of the series variability was accounted for by the chosen predictors. This can adversely affect choice of predictors and hence the validity of the model. The autoregression procedure accounted for first-order autocorrelated residuals and provided reliable estimates of both goodness-of-fit measures and significance levels of chosen predictor variables. The Autoregression procedure by Regression Software "SPSS 10.0.5" has been selected for the curve estimation procedure in present work. Regression model (1.50–1.57) has been built based on field tests data and in the final procedure it has been compared with "MOC" for numerical modeling and simulation results. Computer models are an ideal tool for tracking momentum, inertia, and friction as the transient evolves, and for correctly accounting for changes in mass and energy at boundaries. Note that transients propagate throughout the entire pressurized system. (Figure 1)

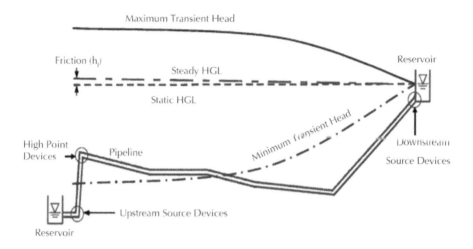

FIGURE 1   Typical locations where transient pulses initiate.

Pulsation: Pulsation generally occurs when a liquid's motive force is generated by reciprocating or peristaltic positive displacement pumps. It is most commonly caused by the acceleration and deceleration of the pumped fluid. This uncontrolled energy appears as pressure spikes.

Vibration and interpenetration between the water flows and mixture components is the visible example of pulsation and is the culprit that usually leads the way to component failure. Pumps: a pump's motor exerts torque on a shaft that delivers energy to the pump's impeller, forcing it to rotate and add energy to the fluid as it passes from the suction to the discharge side of the pump volute. Pumps convey fluid to the downstream end of a system whose profile can be either uphill or downhill, with irregularities such as local high or low points. When the pump starts, pressure can increase rapidly. Whenever power sags or fails, the pump slows or stops and a sudden drop in pressure propagates downstream (a rise in pressure also propagates upstream in the suction system). The similarity of the transient conditions caused by different source devices provides the key to transient analysis in a wide range of different systems: understand the initial state of the system and the ways in which energy and mass are added or removed from it. This is best illustrated by an example for a typical pumping system. (Figures 2–7).

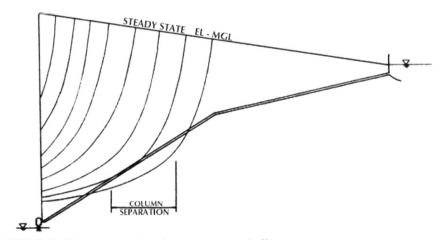

FIGURE 2   Column separations due to pump turned off.

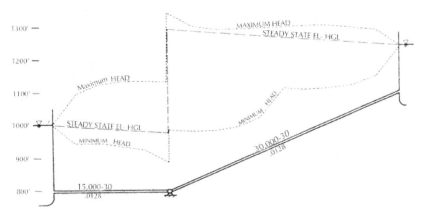

FIGURE 3   Max and min pressure due to pump turned off.

Unlike centrifugal pumps (which produce normally non-damaging high-frequency but low-amplitude pulses), the amplitude is the problem because it is the pressure spike. The peak, instantaneous pressure required to accelerate the liquid in the pipe line can be greater than 10 times the steady state flow pressure produced by a centrifugal pump. Damage to seals gauges diaphragms, valves and joints in piping result from the pressure spikes created by the pulsating flow.

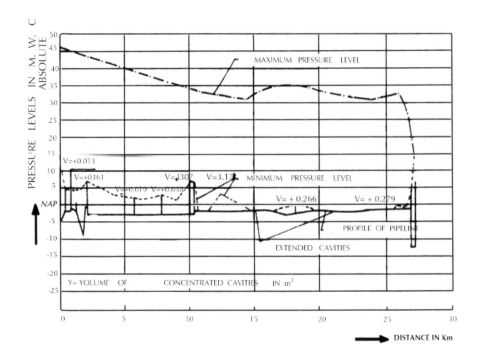

FIGURE 4   Continues and compact cavitations and pulsation in pipe at pump turned off.

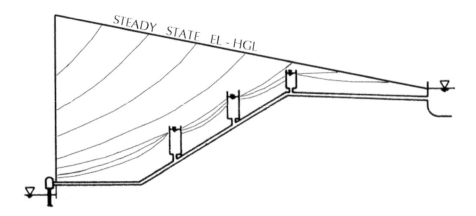

FIGURE 5   Effect of installation of one way surge tank on hydraulic grade line at pump turned off.

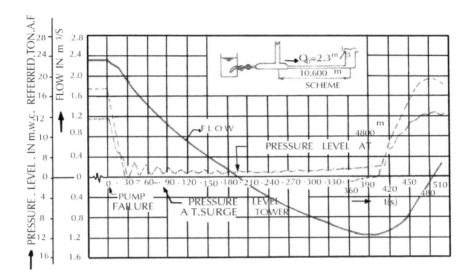

FIGURE 6   Pressure and flow rate changes after pump turned off in system with one way surge tank.

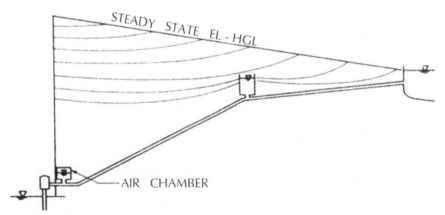

FIGURE 7   Effect of Air chambers and surge tank installation on minus pressure waves motion after pump turned off.

5.3   TIMING AND SHAPE OF TRANSIENT PRESSURE PULSES

With respect to timing, there should be close agreement between the computed and measured periods of the system, regardless of what flow-control operation initiated the transient. With a well-calibrated model of the system, it is possible to use the model in the operational control of the system and anticipate the effects of specific flow-control operations. This requires field measurements to quantify system's pressure-wave speed and friction, with the following considerations: Field measurements can clearly indicate the evolution of the transient. The pressure-wave speed for a pipe with typical material and bedding can be determined if the period of the transient (4 L/a) and the length (L) between measurement locations is known. If there is air in the system, the measured wave speed may be much lower than the theoretical speed. If friction is significant in a system, real-world transients attenuate faster than the numerical simulation, particularly during longer time periods (t > 2 L/a). Poor friction representation does not explain lack of agreement with an initial transient pulse. In general, if model peaks arrive at the wrong time, the wave speed must be adjusted. If model peaks have the wrong shape, the description of the control event (pump shutdown or valve closure) should be adjusted. If the transient dies off too quickly or slowly in the model, the friction losses must be adjusted. If there are secondary peaks, important loops and diversions may need to be included in the model.

5.3.1   Transient Forces

Computations

In accordance with Newton's third law, the force exerted on the piping by the conveyed liquid is equal and opposite to that applied on the liquid by the piping. On physical grounds, the later is due to the following causes: gravity, fluid friction drag, and changes in pipe diameter and/or direction. The linear-momentum and action-reaction principles are applied to an appropriate control volume (CV) to construct general formulae for instantaneous forces applied to pipe walls by the conveyed liquids. Specifically, a fixed CV is defined as being centered on a node, which can be internal (associated with multiple pipes) or external (at the end of exactly one pipe), respectively (Figure 8 and Figure 9).

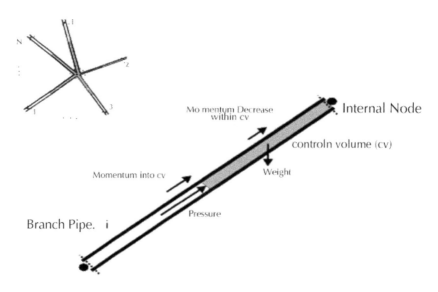

FIGURE 8   Control volumes for internal node.

Analysis of Transient Forces

At zero flow (static or stagnant condition) piping system experiences hydraulic forces due to the weight and static pressure of the liquid to be conveyed. At steady-state, these forces are typically balanced such that forces on most elbows are balanced by forces at another elbow or by a restraint, such as a thrust block. Research refers to this balanced hydraulic steady-state as the "Operating" pressure and temperature.

Pipe stress software can be used to ensure that supports, guides, and restraints are sufficiently strong to hold the pipes in position without excessive displacement or vibration. Hydraulic transients occur whenever a change in flow and/or pressure is rapid with respect to the characteristic time of the system. Rapid changes in pressure and momentum occur during a transient event. This cause liquids and gases to exert transient forces on piping and appurtenances.

This is highly significant for in-plant, buried, and freely-supported piping because, if pressures and flows change during the transient event, the force vectors will likewise change in magnitude and direction. This has fundamental implications for the design of thrust blocks and restraints.

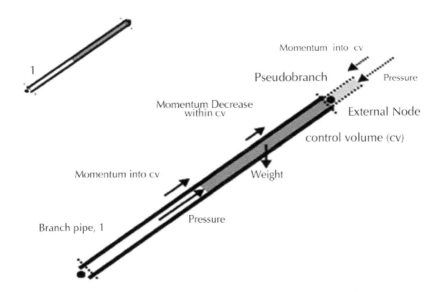

FIGURE 9   Control volumes for external node.

Due to weight, transient forces are always 3D even for horizontal pipelines. For buried piping, these forces are also resisted in 3D at discrete points (thrust blocks), transversely due to contact with the earth, and longitudinally due to pipe friction with the soil.

Transient forces are not linearly proportional to transient pressures. A small increase in transient pressure can develop proportionally larger transient forces. This is because the forces are not a linear function of the pressures.

Thrust blocks or restraints designed for the steady-state or "operating case" times a (constant) safety factor can often be inadequate to resist transient forces, especially for systems with high operating pressures, temperatures or mass.

5.4 EQUATIONS DESCRIBING THE FLUID INTERPENETRATION

The miscible liquids interpenetration happened when they move themselves in the separated pipes toward the common joint and pipe. Fluid condition for example: velocity, pressure, temperature, and the other properties in the two pipes are as similar and the main approach is the changes study on behavior of the fluids flow state. Interpenetration between water flows like the interpenetration of two fluids at parallel between plates and turbulent moving in pipe are common phenomena in our daily life. According to Reynolds number magnitude, separation of fluid direction happened. For fluid motion modeling, 2D-component disperses fluid motion used.

Modeling of two-phase liquid-liquid flows through Kinetics static mixer by means of computational fluid dynamics (CFD) has been presented. The two modeled phases were assumed viscous and Newtonian with the physical properties mimicking an aqueous solution in the continuous and oil in the dispersed (secondary) phase. Three levels of superficial flow velocity were chosen to result in Reynolds numbers equal to 100, 200, and 400, respectively.

The numerical simulations were performed with the help of the commercial software. The modeling involved both block-structured grids and fully non-structured grids for a static mixer with 10 Kinetics inserts. Each of the two grid types had three density levels. The algebraic slip mixture (ASM) model was used in the Eulerian frame of reference and enabled the prediction of the pressure drop across the inserts, the local velocities and volume fraction of the two phases (Figure 10).

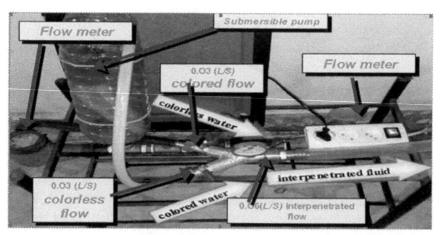

FIGURE 10   Interpenetration and turbulent moving in pipe velocity (u1 and u2) for 2D fluids.

The Lagrangian approach was used to track the trajectory of dispersed fluid elements (drops) in the simulated static mixer. The particle history was analyzed in terms of the residence time in the mixer. While two relaxing miscible fluids are mixed together, their appearances in terms of colors and shapes will change due to their mixing interpenetration (Figure 11).

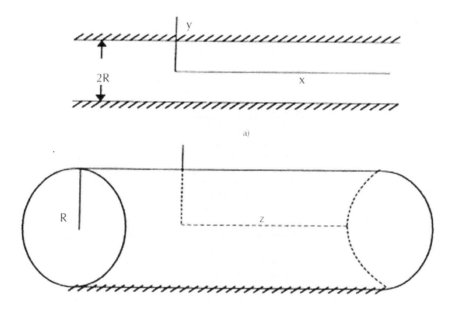

FIGURE 11 Interpenetration of two 2D relaxing fluids at parallel between plates and turbulent moving pipe.

Use equations of motion of two relaxing fluids in pipe are as flowing:

$$u_1 = u_1(y,t) , \quad u_2 = u_2(y,t)$$

$$\left.
\begin{aligned}
\rho_1 \frac{\partial u_1}{\partial t} &= f_1 \mu_1 \frac{\partial^2 u_1}{\partial y^2} + k(u_2 - u_1) - f_1 \frac{\partial p}{\partial x} , \\
\rho_2 \frac{\partial u_2}{\partial t} &= f_2 \mu_2 \frac{\partial^2 u_2}{\partial y^2} + k(u_1 - u_2) - f_2 \frac{\partial p}{\partial x} , \\
\frac{\partial p}{\partial y} &= 0 , \quad \frac{\partial p}{\partial z} = 0 , \quad f_1 + f_2 = 0
\end{aligned}
\right\} \tag{1.1}$$

where, $\bar{u}$, = velocity (m/s), P = pressure, k = module of elasticity of water (kg/m²), f = Darcy–Weisbach friction factor (obtained from Moody diagram) for each pipe, μ = fluid dynamic, viscosity (kg/m.s), ρ = density (kg/m³).

Calculation for equation of motion for relaxing fluids:

$$\left.\begin{array}{c} \theta_1 \dfrac{\partial \tau_1}{\partial t} + \tau_1 = \mu_1 \dfrac{\partial u_1}{\partial y}, \\[2mm] \theta_2 \dfrac{\partial \tau_2}{\partial t} + \tau_2 = \mu_2 \dfrac{\partial u_2}{\partial y} \end{array}\right\} \tag{1.2}$$

θ_1, θ_2 = relaxing time of fluids, define equation of motion for interpenetration of two 2D pressurized relaxing fluids at parallel between plates and turbulent moving in pipe as flowing:

$$\left.\begin{array}{l} \rho_1 \dfrac{\partial u_1}{\partial t} + \rho_1 \theta_1 \dfrac{\partial^2 u_1}{\partial t^2} = f_1 \mu_1 \dfrac{\partial^2 u_1}{\partial y^2} + \theta_1 k \dfrac{\partial (u_1 - u_2)}{\partial t} + k(u_2 - u_1) - f_1\left[\theta_1 \dfrac{\partial^2 p}{\partial t \partial x} + \dfrac{\partial p}{\partial x}\right] \\[3mm] \rho_2 \dfrac{\partial u_2}{\partial t} + \rho_2 \theta_2 \dfrac{\partial^2 u_2}{\partial t^2} = f_2 \mu_2 \dfrac{\partial^2 u_2}{\partial y^2} + \theta_2 k \dfrac{\partial (u_1 - u_2)}{\partial t} + k(u_1 - u_2) - f_2\left[\theta_2 \dfrac{\partial^2 p}{\partial t \partial x} + \dfrac{\partial p}{\partial x}\right] \\[3mm] \dfrac{\partial p}{\partial y} = 0 , \quad \dfrac{\partial p}{\partial z} = 0 , \\[3mm] f_1 + f_2 = 0 \end{array}\right\} \tag{1.3}$$

From (1.3) concluded that pressure drop $\partial p / \partial x$ it is not effective but time is effective.

Assumed that at first time both plan are stopped and pressure at coordination for this time is low.

$$\left.\begin{array}{l} t = 0 \begin{cases} u_1 = 0 , \ u_2 = 0 \\ \partial u_1 / \partial t = 0 , \ \partial u_2 / \partial t = 0 \end{cases} \\[3mm] y = h \quad (t > 0) \quad u_1 = 0 \ \ u_2 = 0 \\[1mm] y = -h \quad (t > 0) \quad u_1 = 0 \ \ u_2 = 0 \end{array}\right\} \tag{1.4}$$

At time t condition with Laplace rule, with Equations (1.3) and (1.4) we have:

$$\left.\begin{array}{l} \dfrac{d^2 \bar{u}_1}{dy^2} - \alpha_1 \bar{u}_1 + \beta_1 \bar{u}_2 = \dfrac{1}{\mu_1}\dfrac{\partial P}{\partial x} \\[3mm] \dfrac{d^2 \bar{u}_2}{dy^2} - \alpha_2 \bar{u}_2 + \beta_2 \bar{u}_1 = \dfrac{1}{\mu_2}\dfrac{\partial P}{\partial x} \end{array}\right\} \tag{1.5}$$

with:

$$\left.\begin{array}{l} y = h \quad \bar{u}_1 = 0 , \ \bar{u}_2 = 0 \\ y = -h \quad \bar{u}_1 = 0 , \ \bar{u}_2 = 0 \end{array}\right\} \tag{1.6}$$

where:

$$
\left.\begin{array}{l}
\alpha_1 = \dfrac{\rho_1(\theta_1 s^2 + s) + k(\theta_1 s + 1)}{f_1 \mu_1} \ , \\[2mm]
\beta_1 = \dfrac{k(\theta_1 s + 1)}{f_1 \mu_1} \ , \\[2mm]
\alpha_2 = \dfrac{\rho_2(\theta_2 s^2 + s) + k(\theta_2 s + 1)}{f_2 \mu_2} \ , \\[2mm]
\beta_2 = \dfrac{k(\theta_2 s + 1)}{f_2 \mu_2} \ ,
\end{array}\right\}
\qquad (1.7)
$$

Calculation $\partial p / \partial x = A = const$ and with product of Equation (1.5) into N flowing differential equation received:

$$
\frac{d^2}{dy^2}(N\overline{u}_1 + \overline{u}_2) - (\alpha_2 - N\beta_1)(N\overline{u}_1 + \overline{u}_2) = \left[\frac{N(1 + \theta_1 s)}{\mu_1} + \frac{1 + \theta_2 s}{\mu_2}\right] A \ , \qquad (1.8)
$$

$$
N_{1,2} = \frac{-(\alpha_1 - \alpha_2) \pm \sqrt{(\alpha_1 - \alpha_2)^2 + 4\beta_1 \beta_2}}{2\beta_1} \ .
$$

Equation (1.6) calculated with Equation (1.8):

$$
A\overline{u}_1 + \overline{u}_2 = -A\left[\frac{N}{\mu_1} + \frac{1}{\mu_2}\right]\left[1 - \frac{ch\sqrt{\alpha_2 - N\beta_1}\, y}{ch\sqrt{\alpha_2 - N\beta_1}\, h}\right] .
$$

N calculation with two meaning:

$$
N_1 \overline{u}_1 + \overline{u}_2 = -A\left[\frac{N_1}{\mu_1} + \frac{1}{\mu_2}\right]\left[1 - \frac{ch\sqrt{\alpha_2 - N_1 \beta_1}\, y}{ch\sqrt{\alpha_2 - N_1 \beta_1}\, h}\right] ,
$$

$$
N_2 \overline{u}_1 + \overline{u}_2 = -A\left[\frac{N_2}{\mu_1} + \frac{1}{\mu_2}\right]\left[1 - \frac{ch\sqrt{\alpha_2 - N_2 \beta_1}\, y}{ch\sqrt{\alpha_2 - N_2 \beta_1}\, h}\right] ,
$$

where for equation velocity find:

$$
\overline{u}_1 = \frac{A}{N_2 - N_1}\left\{\frac{\frac{N_1}{\mu_1} + \frac{1}{\mu_2}}{\alpha_2 - N_1 \beta_1}\left[1 - \frac{ch\sqrt{\alpha_2 - N_1 \beta_1}\, y}{ch\sqrt{\alpha_2 - N_1 \beta_1}\, h}\right] - \frac{\frac{N_2}{\mu_1} + \frac{1}{\mu_2}}{\alpha_2 - N_2 \beta_1}\left[1 - \frac{ch\sqrt{\alpha_2 - N_2 \beta_1}\, y}{ch\sqrt{\alpha_2 - N_2 \beta_1}\, h}\right]\right\} ,
$$

$$
\overline{u}_2 = \frac{A}{N_1 - N_2}\left\{\frac{\frac{N_2}{\mu_1} - \frac{\beta_2}{\beta_1}\frac{1}{\mu_2}}{\alpha_2 - N_1 \beta_1}\left[1 - \frac{ch\sqrt{\alpha_2 - N_1 \beta_1}\, y}{ch\sqrt{\alpha_2 - N_1 \beta_1}\, h}\right] - \frac{\frac{N_1}{\mu_2} - \frac{\beta_2}{\beta_1}\frac{1}{\mu_1}}{\alpha_2 - N_2 \beta_1}\left[1 - \frac{ch\sqrt{\alpha_2 - N_2 \beta_1}\, y}{ch\sqrt{\alpha_2 - N_2 \beta_1}\, h}\right]\right\} .
$$

$$\bar{u}_i = \frac{1}{2\pi i}\int_{\sigma-i\yen}^{\sigma+i\yen}\frac{A}{N_2-N_1}\left\{\frac{\dfrac{N_1}{\mu_1}-\dfrac{1}{\mu_2}}{\alpha_2-N_1\beta_1}\left[1-\frac{ch\sqrt{\alpha_2-N_1\beta_1}\,y}{ch\sqrt{\alpha_2-N_1\beta_1}\,h}\right]-\right.$$

$$\left.-\frac{\dfrac{N_2}{\mu_1}-\dfrac{1}{\mu_2}}{\alpha_2-N_2\beta_1}\left[1-\frac{ch\sqrt{\alpha_2-N_2\beta_1}\,y}{ch\sqrt{\alpha_2-N_2\beta_1}\,h}\right]\right\}.e^{st}\frac{ds}{s}$$

In this calculation we have:

$$s=s_1\quad,\quad s=s_2\quad,\quad s=s_3\quad,\quad s=s_4 \ :s_1,s_2,s_3,s_4,$$

$$s_{1n}=\gamma_{1n}\quad,\quad s_{2n}=\gamma_{2n}\quad,\quad s_{3n}=\gamma_{3n}\quad,\quad s_{4n}=\gamma_{4n}\quad,\quad \gamma_{in}$$

Proportional to forth procedure:

$$\alpha_2-N_1\beta_1=-\frac{\pi^2}{h^2}\left(n+\frac{1}{2}\right)^2,$$

$$\alpha_2-N_2\beta_1=-\frac{\pi^2}{h^2}\left(n+\frac{1}{2}\right)^2,$$

In this state for velocity we have:

$$u_1=-\frac{\dfrac{1}{f_1\mu_1 f_2\mu_2}}{\dfrac{1}{f_1\mu_1}+\dfrac{1}{f_2\mu_2}}A\left\{-\frac{1}{2}(y^2-h^2)+\frac{\left(\dfrac{1}{\mu_1}-\dfrac{1}{\mu_2}\right)f_2\mu_2}{\left(\dfrac{1}{f_1\mu_1}+\dfrac{1}{f_2\mu_2}\right)k}\times\right.$$

$$\left.\times\left[1-\frac{ch\sqrt{\left(\dfrac{1}{f_1\mu_1}+\dfrac{1}{f_2\mu_2}\right)ky}}{ch\sqrt{\left(\dfrac{1}{f_1\mu_1}+\dfrac{1}{f_2\mu_2}\right)kh}}\right]\right\}+\frac{4A}{\pi}\sum_{i=1}^{4}\sum_{n=1}^{\infty}\frac{(-1)^{n+1}}{\left(n+\dfrac{1}{2}\right)}\cos\left[\pi\left(n+\frac{1}{2}\right)\frac{y}{h}\right]\times$$

$$\times \left\{ \frac{\left[\frac{\pi^2}{h^2}\left(n+\frac{1}{2}\right) + \frac{\left(\theta^2\gamma_{in}+1\right)\left(\rho^2\gamma_{in}+k\right)}{f_2\mu_2} \right]\frac{1}{\mu_1} + \frac{k\left(\theta_1\gamma_{in}+1\right)}{f_1\mu_1}\frac{1}{\mu_2}}{2\gamma_{in}\left(\frac{\theta_1\rho_1}{f_1\mu_1}+\frac{\theta_2\rho_2}{f_2\mu_2}\right)+\frac{\theta_1k+\rho_1}{f_1\mu_1}+\frac{\theta_2k+\rho_2}{f_2\mu_2}+\frac{\frac{\left(\theta_1\gamma_{in}+1\right)\left(\rho_1\gamma_{in}+k\right)}{f_1\mu_1}}{2\frac{\pi^2}{h^2}\left(n+\frac{1}{2}\right)^2+\frac{\left(\theta_1\gamma_{in}+1\right)\left(\rho_1\gamma_{in}+k\right)}{f_1\mu_1}}} \cdot \frac{\frac{\left(\theta_1\gamma_{in}+1\right)\left(\rho_1\gamma_{in}+k\right)}{f_1\mu_1}+\frac{\left(\theta_2\gamma_{in}+1\right)\left(\rho_2\gamma_{in}+k\right)}{f_2\mu_2}-2\frac{\pi^2}{h^2}\left(n+\frac{1}{2}\right)^2}{1}^{\frac{1}{\gamma_{in}}} \right.$$

$$\left. \frac{e^{-\gamma_{in}t}}{-\frac{\left(\theta_2\gamma_{in}+1\right)\left(\rho_2\gamma_{in}+k\right)}{f_2\mu_2}\left(2\gamma_{in}\left(\frac{\theta_1\rho_1}{f_1\mu_1}+\frac{\theta_2\rho_2}{f_2\mu_2}\right)-\frac{\theta_2k+\rho_2}{f_2\mu_2}-\frac{\theta_1k+\rho_1}{f_1\mu_1}\right)}{\left(\frac{\left(\theta_2\gamma_{in}+1\right)\left(\rho_2\gamma_{in}+k\right)}{f_2\mu_2}\right)} \right\}, \quad (1.9)$$

$$u_2 = -\frac{\frac{1}{f_1\mu_1 f_2\mu_2}}{\frac{1}{f_1\mu_1}+\frac{1}{f_2\mu_2}} A\left\{ -\frac{1}{2}\left(y^2+h^2\right) + \frac{\left(\frac{1}{\mu_1}-\frac{1}{\mu_2}\right)f_1\mu_1}{\left(\frac{1}{f_1\mu_1}-\frac{1}{f_2\mu_2}\right)k} \times \right.$$

$$\left. \times \left[1-\frac{ch\sqrt{\left(\frac{1}{f_1\mu_1}+\frac{1}{f_2\mu_2}\right)ky}}{ch\sqrt{\left(\frac{1}{f_1\mu_1}+\frac{1}{f_2\mu_2}\right)kh}}\right]\right\} + \frac{4A}{\pi}\sum_{i=1}^{4}\sum_{n=1}^{\infty}\frac{(-1)^{n+1}}{\left(n+\frac{1}{2}\right)}\cos\left[\pi\left(n+\frac{1}{2}\right)\frac{y}{h}\right] \times$$

$$\times \left\{ \frac{\left| \frac{\pi^2}{h^2}\left(n+\frac{1}{2}\right) + \frac{\left(\theta^2\gamma_{in}+1\right)\left(\rho_1\gamma_{in}+k\right)}{f_1\mu_1}\right| \frac{1}{\mu_2} + \frac{k\left(\theta_1\gamma_{in}+1\right)}{f_2\mu_2}\frac{1}{\mu_1}}{\frac{\left(\theta_1\gamma_{in}+1\right)\left(\rho_1\gamma_{in}+k\right)}{f_1\mu_1} + \frac{\left(\theta_2\gamma_{in}+1\right)\left(\rho_2\gamma_{in}+k\right)}{f_2\mu_2} - 2\frac{\pi^2}{h^2}\left(n+\frac{1}{2}\right)^2}\cdot\frac{1}{\gamma_{in}} }{2\gamma_{in}\left(\frac{\theta_1\rho_1}{f_1\mu_1}+\frac{\theta_2\rho_2}{f_2\mu_2}\right)+\frac{\theta_1 k+\rho_1}{f_1\mu_1}+\frac{\theta_2 k+\rho_2}{f_2\mu_2}+\frac{\left(\frac{\left(\theta_1\gamma_{in}+1\right)\left(\rho_1\gamma_{in}+k\right)}{f_1\mu_1}\right)}{2\frac{\pi^2}{h^2}\left(n+\frac{1}{2}\right)^2+\frac{\left(\theta_1\gamma_{in}+1\right)\left(\rho_1\gamma_{in}+k\right)}{f_1\mu_1}}} \right.$$

$$\left. \frac{e^{-\gamma_{in}t}}{-\frac{\left(\theta_2\gamma_{in}+1\right)\left(\rho_2\gamma_{in}+k\right)}{f_2\mu_2}}\left|\left|2\gamma_{in}\left(\frac{\theta_1\rho_1}{f_1\mu_1}+\frac{\theta_2\rho_2}{f_2\mu_2}\right)-\frac{\theta_2 k+\rho_2}{f_2\mu_2}-\frac{\theta_1 k+\rho_1}{f_1\mu_1}+\frac{4k^2\left(\theta_1\theta_2\gamma_{in}+\theta_1+\theta_2\right)}{f_1\mu_1 f_2\mu_2}\right.\right.}{+\frac{\left(\theta_1\gamma_{in}+1\right)\left(\rho_1\gamma_{in}+k\right)}{f_1\mu_1}-\frac{\left(\theta_2\gamma_{in}+1\right)\left(\rho_2\gamma_{in}+k\right)}{f_2\mu_2}} \right\}, \quad (1.10)$$

When $\theta_1 = \theta_2 = 0$ from Equations (1.9) and (1.10) we have:

$$\theta_1 = \theta_2 = 0$$

$$\mu_1 = \mu_2 \quad , \quad \rho_{1i} = \rho_{2i} \,,$$

$$u = u_1 = u_2 = \frac{A}{2\mu}\left(h^2 - y^2\right) - \frac{16h^2 A}{\pi\mu}\sum_{n=1}^{\infty}\frac{(-1)^n}{(2n+1)^3}\cos\frac{(2n+1)}{2h}y.e^{-\frac{\pi^2}{h^2}\left(n+\frac{1}{2}\right)^2\frac{\mu}{\rho}t}$$

At condition $t \to \infty$ for unsteady motion of fluid, it is easy for calculation table procedure:

$$\left.\begin{aligned} \rho_1\frac{\partial u_1}{\partial t} &= f_1\mu_1\left(\frac{\partial^2 u_1}{\partial r^2}+\frac{1}{r}\frac{\partial u_1}{\partial r}\right)+k(u_2-u_1)-f_1\frac{\partial P}{\partial z} \\ \rho_2\frac{\partial u_2}{\partial t} &= f_2\mu_2\left(\frac{\partial^2 u_2}{\partial r^2}+\frac{1}{r}\frac{\partial u_2}{\partial r}\right)+k(u_1-u_2)-f_2\frac{\partial P}{\partial z} \end{aligned}\right\} \quad (1.11)$$

For every relaxing phase we have:

$$\left. \begin{array}{l} \theta_1 \dfrac{\partial \tau_1}{\partial t} + \tau_1 = \mu_1 \dfrac{\partial u_1}{\partial r} , \\[3mm] \theta_2 \dfrac{\partial \tau_2}{\partial t} + \tau_2 = \mu_2 \dfrac{\partial u_2}{\partial r} , \end{array} \right\} \tag{1.12}$$

Start and limiting conditions:

$$\begin{array}{ll} t = 0 & u_1 = 0 , \ u_2 = 0 , \\[2mm] r = R(t > 0) & u_1 = 0 , \ u_2 = 0 . \end{array} \tag{1.13}$$

In condition of differential Equation (1.11) by $\partial \tau_1/\partial t$ from Equation (1.12) and with τ_1 concluded:

$$\left. \begin{array}{l} \rho_1 \left(\dfrac{\partial u_1}{\partial t} + \theta_1 \dfrac{\partial^2 u_1}{\partial t^2} \right) = f_1 \mu_1 \left(\dfrac{\partial^2 u_1}{\partial t^2} + \dfrac{1}{r} \dfrac{\partial u_1}{\partial r} \right) + k \left[\theta_1 \dfrac{\partial}{\partial t} (u_2 - u_1) + (u_2 - u_1) \right] - \\[4mm] - f_1 \left[\theta_1 \dfrac{\partial^2 p}{\partial t \partial z} + \dfrac{\partial p}{\partial z} \right] , \\[4mm] \rho_2 \left(\dfrac{\partial u_2}{\partial t} + \theta_2 \dfrac{\partial^2 u_2}{\partial t^2} \right) = f_2 \mu_2 \left(\dfrac{\partial^2 u_2}{\partial r^2} + \dfrac{1}{r} \dfrac{\partial u_2}{\partial r} \right) + k \left[\theta_2 \dfrac{\partial}{\partial t} (u_1 - u_2) + (u_1 - u_2) \right] - \\[4mm] - f_2 \left[\theta_2 \dfrac{\partial^2 p}{\partial t \partial z} + \dfrac{\partial p}{\partial z} \right] . \end{array} \right\} \tag{1.14}$$

Data condition (1.13) and integration. In this condition Laplace is toward Equation (1.14). Then solution find in the form of velocity equation, 1D fluid viscosity in round pipe is:

$$u_1 = -\dfrac{A}{f_1 \mu_1 \left(\dfrac{1}{f_1 \mu_1} + \dfrac{1}{f_2 \mu_2} \right)} \left\{ \dfrac{1}{4} (r^2 - R^2) \dfrac{1}{f_2 \mu_2} + \dfrac{\left(\dfrac{1}{\mu_2} - \dfrac{1}{\mu_1} f_2 \mu_2 \right)}{k \left(\dfrac{1}{f_1 \mu_1} + \dfrac{1}{f_2 \mu_2} \right)} \times \right.$$

$$\left. \times \left[1 - \dfrac{I_0 \left(\sqrt{\dfrac{1}{f_1 \mu_1} + \dfrac{1}{f_2 \mu_2}} \, kr \right)}{I_0 \left(\sqrt{\dfrac{1}{f_1 \mu_1} + \dfrac{1}{f_2 \mu_2}} \, kR \right)} \right] \right\} + \sum_{i=1}^{4} \sum_{n=1}^{\infty} \dfrac{4A}{\alpha_n} \dfrac{J_0 \left(\alpha_n \dfrac{r}{R} \right)}{J_1 (\alpha_n)} \times$$

$$\times\left\{\frac{\dfrac{\alpha_n^2}{R^2}+\dfrac{(\theta_2\gamma_{in}+1)(-\rho_2\gamma_{in}+k)}{f_1\mu_1}\dfrac{1}{\mu_1}+\dfrac{k(\theta_1\gamma_{in}+1)}{f_1\mu_1}\dfrac{1}{\mu_2}}{\dfrac{(\theta_1\gamma_{in}+1)(-\rho_1\gamma_{in}+k)}{f_1\mu_1}+\dfrac{(\theta_2\gamma_{in}+1)(-\rho_2\gamma_{in}+k)}{f_2\mu_2}\pm 2\dfrac{\alpha_n^2}{R^2}}\times\right.$$

$$\times\frac{e^{-\gamma_{in}t}}{\gamma_{in}}\bigg/\left\{2\gamma_{in}\left(\frac{\theta_1\rho_1}{f_1\mu_1}+\frac{\theta_2\rho_2}{f_2\mu_2}\right)\pm\frac{\theta_1 k+\rho_1}{f_1\mu_1}+\frac{\theta_2 k+\rho_2}{f_2\mu_2}+\right.$$

$$+\left[\left(\frac{(\theta_1\gamma_{in}+1)(\rho_1\gamma_{in}+k)}{f_1\mu_1}-\frac{(\theta_2\gamma_{in}+1)(\rho_2\gamma_{in}+k)}{f_2\mu_2}\right)\left(2\gamma_{in}\left(\frac{\theta_1\rho_1}{f_1\mu_1}+\frac{\theta_2\rho_2}{f_2\mu_2}\right)-\right.\right.$$

$$\left.\left.-\frac{\theta_2 k+\rho_2}{f_2\mu_2}-\frac{\theta_1 k+\rho_1}{f_1\mu_1}+4k^2\frac{\theta_1\theta_2\gamma_{in}+\theta_1+\theta_2}{f_1\mu_1 f_2\mu_2}\right)\right]\bigg/\left[2\frac{\alpha_n^2}{R}+\right.$$

$$\left.\left.+\frac{(\theta_1\gamma_{in}+1)(-\rho_1\gamma_{in}+k)}{f_1\mu_1}+\frac{(\theta_2\gamma_{in}+1)(-\rho_2\gamma_{in}+k)}{f_2\mu_2}\right]\right\}, \qquad (1.15)$$

$$u_2=-\frac{A}{f_1\mu_1\left(\dfrac{1}{f_1\mu_1}+\dfrac{1}{f_2\mu_2}\right)}\left\{\frac{1}{4}(r^2-R^2)\frac{1}{f_2\mu_2}+\frac{\left(\dfrac{1}{\mu_2}-\dfrac{1}{\mu_1}\right)f_2\mu_2}{k\left(\dfrac{1}{f_1\mu_1}+\dfrac{1}{f_2\mu_2}\right)}\times\right.$$

$$\times\left[1-\frac{I_0\sqrt{\left(\dfrac{1}{f_1\mu_1}+\dfrac{1}{f_2\mu_2}\right)}kr}{I_0\sqrt{\left(\dfrac{1}{f_1\mu_1}+\dfrac{1}{f_2\mu_2}\right)}kR}\right]+\sum_{i=1}^{4}\sum_{n=1}^{\infty}\frac{4A}{\alpha_n}\frac{J_0\left(\alpha_n\dfrac{r}{R}\right)}{J_1(\alpha_n)}\times$$

$$\times\left\{\frac{\dfrac{\alpha_n^2}{R^2}+\dfrac{(\theta_2\gamma_{in}+1)(\rho_1\gamma_{in}+k)}{f_1\mu_1}\dfrac{1}{\mu_2}+\dfrac{k(\theta_2\gamma_{in}+1)}{f_2\mu_2}\dfrac{1}{\mu_1}}{\dfrac{(\theta_1\gamma_{in}+1)(-\rho_1\gamma_{in}+k)}{f_1\mu_1}+\dfrac{(\theta_2\gamma_{in}+1)(-\rho_2\gamma_{in}+k)}{f_2\mu_2}\pm 2\dfrac{\alpha_n^2}{R^2}}\times\right.$$

$$\times \frac{e^{-\gamma_{in}t}}{\gamma_{in}} \Bigg/ \left\{ 2\gamma_{in}\left(\frac{\theta_1\rho_1}{f_1\mu_1} + \frac{\theta_2\rho_2}{f_2\mu_2}\right) + \frac{\theta_1 k + \rho_1}{f_1\mu_1} + \frac{\theta_2 k - \rho_2}{f_2\mu_2} \pm \right.$$

$$\times \frac{e^{-\gamma_{in}t}}{\gamma_{in}} \Bigg/ \left\{ 2\gamma_{in}\left(\frac{\theta_1\rho_1}{f_1\mu_1} + \frac{\theta_2\rho_2}{f_2\mu_2}\right) + \frac{\theta_1 k + \rho_1}{f_1\mu_1} + \frac{\theta_2 k - \rho_2}{f_2\mu_2} \pm \right.$$

$$\pm \left[\left(\frac{(\theta_1\gamma_{in}+1)(\rho_1\gamma_{in}+k)}{f_1\mu_1} - \frac{(\theta_2\gamma_{in}+1)}{f_2\mu_2}\right) \left(2\gamma_{in}\left(\frac{\theta_1\rho_1}{f_1\mu_1} + \frac{\theta_2\rho_2}{f_2\mu_2}\right) + \right.\right.$$

$$+ \frac{\theta_2 k + \rho_2}{f_2\mu_2} - \frac{\theta_1 k + \rho_1}{f_1\mu_1} + 4k^2 \frac{\theta_1\theta_2\gamma_{in} + \theta_1 + \theta_2}{f_1\mu_1 f_2\mu_2}\right) \Bigg] \Bigg/ \left[-2\frac{\alpha_n^2}{R} + \right.$$

$$\pm \left[\frac{(\theta_1\gamma_{in}+1)(\rho_1\gamma_{in}+k)}{f_1\mu_1} - \frac{(\theta_2\gamma_{in}+1)(\rho_2\gamma_{in}+k)}{f_2\mu_2} \right], \quad (1.16)$$

When $\theta_1 = \theta_2 = 0$ from Equations (1.15) and (1.16) we have (1.4) in condition:

$$\begin{cases} \theta_1 = & \theta_2 = & 0 \\ \mu_1 = & \mu_2 = & \mu \\ \rho_{1i} = & \rho_{2i} = & \rho \end{cases}$$

For example, if a valve is closed suddenly at the end of a pipeline system a water hammer wave propagates in the pipe. Moving water in a pipe has kinetic energy proportional to the mass of the water in a given volume times the square of the velocity of the water.

Changes in fluid properties: such as depressurization due to the sudden opening of a relief valve, a propagating pressure pulse, heating, or cooling in cogeneration or industrial systems, mixing with solids or other liquids (may affect fluid density, specific gravity, and viscosity), formation and collapse of vapor bubbles (cavitations), and air entrainment or release from the system (at air vents and/or due to pressure waves).

Changes at system boundaries: such as rapidly opening or closing a valve, pipe burst (due to high pressure) or pipe collapse (due to low pressure), pump start/shift/

stop, air intake at a vacuum breaker, water intake at a valve, mass outflow at a pressure-relief valve or fire hose, breakage of a rupture disk, and hunting and/or resonance at a control valve.

Rigid-column theory describes unsteady flow of an incompressible liquid in a rigid system. It is only applicable to slower transient phenomena. Both branches of transient theory stem from the same governing equations. Research uses the more advanced elastic theory system wide for virtually every simulation, but it can also switch to the faster rigid-column theory (in specific reaches and for special applications) to reduce execution time, as discussed in rigid-column simulation. Rigid-column theory describes unsteady flow of an incompressible liquid in a rigid system. It is only applicable to slower transient phenomena.

Elastic theory describes unsteady flow of a compressible liquid in an elastic system (e.g., where pipes can expand and contract).

$$H_2 - H_1 = (C/g)(V_2 - V_1) = \rho C(V_2 - V_1), \tag{1.17}$$

(Joukowski Formula)

where, $H_2 - H_1$ = pressure difference (m-H_2O), C = surge wave velocity (m/s), g = acceleration of gravity (m/s²), $V_2 - V_1$ = velocity difference (m/s).

$$C = 1/\left[\rho\left((1/k) + (dC_1/Ee)\right)\right]1/2, \tag{1.18}$$

(Allievi Formula)

where, d = pipe diameter (m), e = pipe thickness (m), E = module of elasticity (kg/m²), ρ = density (kg/m³), k = module of elasticity of water (kg/m²), C = wave velocity (m/s), C_1 = pipe support coefficient.

The continuity equation and the momentum equation are needed to determine V and p in a 1D flow system. By olving these two equations produces a theoretical result that usually corresponds quite closely to actual system measurements if the data and assumptions used to build the numerical model are valid. Transient analysis results that are not comparable with actual system measurements are generally caused by inappropriate system data (especially boundary conditions) and inappropriate assumptions. Several factors can contribute to water hammer. Improperly sized piping in relation to water flow velocity, high water pressure with no pressure-reducing valve, straight runs that are too long without bends: Poor strapping of piping system to structure, no dampening system in place to reduce or absorb shockwaves. For example, if a valve is closed suddenly at the end of a pipeline system a water hammer wave propagates in the pipe.

5.5 METHOD OF CHARACTERISTIC SOLUTION FOR PARTIAL DIFFERENTIAL EQUATION

The MOC is the most widely used and tested approach, with support for complex boundary conditions and friction and vaporous cavitations models. Partial differential equation (PDE) of continuity and momentum (e.g., Navier–Stokes) into ordinary differential equations that are solved algebraically along lines called characteristics. The MOC solution is exact along characteristics, but friction, vaporous cavitations and some boundary representations introduce errors in the results.

Newton second law and fluid fine element forces analysis: (Figures 12–16)

$$-(1/\gamma)\,(\partial p/\partial s)-(\partial z/\partial s)-(4\tau/\gamma D)=(1/g)\,(dv/dt) \qquad (1.19)$$

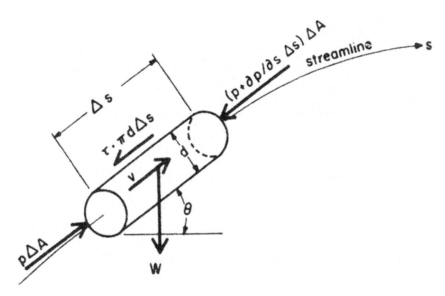

FIGURE 12   Newton second law-forces analysis for fluid element.

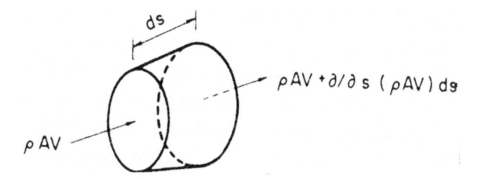

FIGURE 13   Continuity equation or conservation of momentum.

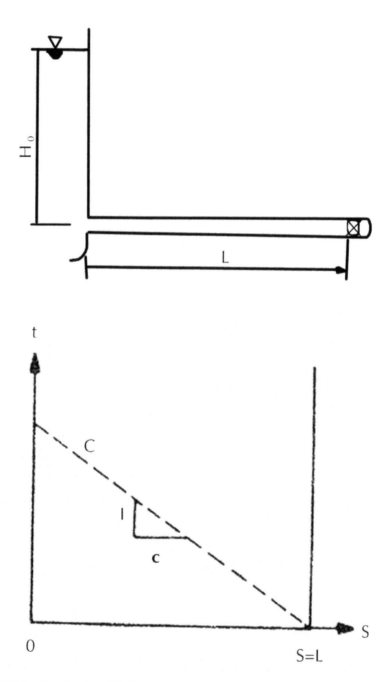

FIGURE 14   Coordination of (s-t).

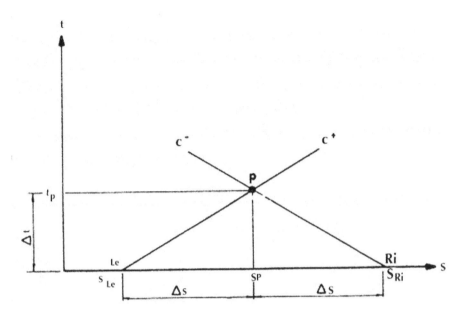

FIGURE 15   Coordination of (s-t) for finding P and V.

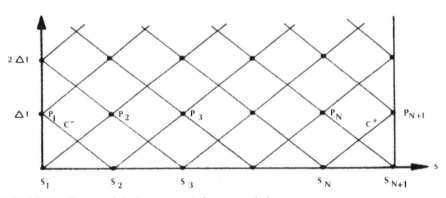

FIGURE 16   Characteristic lines network for assumed pipe.

Equation (1.19) cannot be solved analytically, but they can be expressed graphically in space-time as characteristic lines (or curves), called characteristics, that represent signals propagating to the right (C+) and to the left (C−) simultaneously and from each location in the system. At each interior solution point, signals arrive from the two adjacent points simultaneously. A linear combination of H and V is invariant along each characteristic if friction losses are neglected; therefore, H and V can be obtained exactly at solution points.

With head losses concentrated at solution points and the assumption that friction is small, an iterative procedure is used in conjunction with MOC to advance the solution in time. Transient modeling essentially consists of solving these equations, for every solution point and time step, for a wide variety of boundary conditions and system topologies.

To obtain a general computer model, the following additional capabilities are required:

(1) Boundary conditions must also be expressed as algebraic and/or differential equations based on their physical properties. This must be done for every hydraulic element in the model and solved along with the characteristic equations.

(2) Equations of state are incorporated to model vaporous cavitations, whereby the fluid can flash into vapor at low pressures.

(3) The length of computational reaches must be set to achieve sufficient accuracy without resulting in too small a time step and an excessively long execution time.

(4) Friction losses are assumed to be concentrated at solution points. Different models can be implemented, ranging from steady-state to quasi-steady to unsteady (transient) friction.

(5) Method of Characteristics MOC: is the most widely used and tested approach.

(6) Laboratory Models: a scale model can be built to reproduce transients observed in a prototype (real) system, typically for forensic or steam system investigations. As a design method, this approach is limited by model scale effects and by very high costs. However, models have provided invaluable basic research data on vaporous cavitations and vortex shedding and transient friction.

(7) Field Tests: can provide key modeling parameters such as the pressure-wave speed or pump inertia. Advanced flow and pressure sensors equipped with high-speed data loggers and "PLC" in water pipeline makes it possible to capture fast transients, down to 5 ms. Methods such as inverse transient calibration and leak detection in calculation of unaccounted for water "UFW" use such data [20].

KEYWORDS

- **Algebraic slip mixture**
- **Computational fluid dynamics**
- **Fluid interpenetration**
- **Method of characteristics**
- **Partial differential equation**
- **Regression model**

REFERENCES

1. Joukowski, N. Paper to Polytechnic Soc. Moscow, spring of 1898, English translation by Miss O. *Simin. Proc. AWWA*, 57–58 (1904).
2. Allievi, L. General Theory of Pressure Variation in Pipes. *Ann. D. Ing.*, 166–171 (1982).
3. Wood, F. M. *History of Water hammer*. Civil Engineering Research Report no 65, Queens University, Canada, pp. 66–70 (1970).
4. Parmakian, J. Water Hammer Design Criteria. *J. of Power Div., ASCE*, pp. 456–460 (September, 1957).
5. Parmakian, J. *Water Hammer Analysis*. Dover Publications, Inc., New York, pp. 51–58 (1963).
6. Streeter, V. L. and Lai, C. Water Hammer Analysis Including Fluid Friction. *Journal of Hydraulics Division, ASCE*, **88**, 79 (1962).
7. Streeter, V. L. and Wylie, E. B. *Fluid Mechanics*. McGraw-Hill Ltd., USA, pp. 492–505 (1979).
8. Streeter, V. L. and Wylie, E. B. *Fluid Mechanics*. McGraw-Hill Ltd., USA, pp. 398–420 (1981).
9. Wylie, E. B., Streeter, V. L., and Talpin, L. B. Matched impedance to control fluid transients. *Trans. ASME*, **105**(2), 219–224 (1983).
10. Wylie, E. B. and Streeter, V. L. *Fluid Transients in Systems*. Prentice-Hall, Englewood Cliffs, New Jersey p. 4 (1993).
11. Wylie, E. B. and Streeter, V. L. *Fluid Transients*. Feb Press, Ann Arbor, MI 1982, p. 158 (corrected copy, 1983).
12. Brunone, B., Karney, B. W., Mecarelli, M., and Ferrante, M. Velocity Profiles and Unsteady Pipe Friction in Transient Flow. *Journal of Water Resources Planning and Management, ASCE*, **126**(4), 236–244 (July, 2000).
13. Koelle, E., Luvizotto, Jr. E., and Andrade, J. P. G. *Personality Investigation of Hydraulic Networks using MOC – Method of Characteristics*. Proceedings of the 7th International Conference on Pressure Surges and Fluid Transients, Harrogate Durham, United Kingdom, pp. 1–8 (1996),
14. Filion, Y. and Karney, B. W. *A Numerical Exploration of Transient Decay Mechanisms in Water Distribution Systems*. Proceedings of the ASCE Environmental Water Resources Institute Conference, American Society of Civil Engineers, Roanoke, Virginia, p. 30 (2002).
15. Hamam, M. A. and Mc Corquodale, J. A. Transient Conditions in the Transition from Gravity to Surcharged Sewer Flow. *Canadian J. of Civil Eng.*, 65–98 (Canada) (September, 1982).
16. Savic, D. A. and Walters, G. A. Genetic Algorithms Techniques for Calibrating Network Models, Report No. 95/12. *Centre for Systems and Control Engineering*, 137–146 (1995).
17. Savic, D. A. and Walters, G. A. *Genetic Algorithms Techniques for Calibrating Network Models*. University of Exeter, Exeter, United Kingdom, pp. 41–77 (1995).
18. Walski, T. M. and Lutes, T. L. Hydraulic Transients Cause Low-pressure Problems. *Journal of the American Water Works Association*, **75**(2), 58 (1994).
19. Wu, Z. Y. and Simpson, A. R. Competent genetic-evolutionary optimization of water distribution systems. *J. Comput. Civ. Eng.*, **15**(2), 89–101 (2001).
20. Hariri Asli, K. *Water Hammer Research: Advances in Nonlinear Dynamics Modeling*. PhD. Thesis of Kaveh Hariri Asli, Toronto, Canada, Published by Apple Academic Press, Inc., Exclusive worldwide distribution by CRC Press, a Taylor and Francis Group, Print ISBN: 9781926895314, eBook: 978-1-46-656887-7 (2012), Retrieved from www.AppleAcademicPress.com.

6 Heat Flow and Transient Heat

CONTENTS

6.1 INTRODUCTION

The drying is a complex process involving simultaneous coupled transient heat, mass, and momentum transport. It is a process whereby the moisture is vaporized and swept away from the surface, sometimes in vacuum but normally by means of a carrier fluid passing through or over moist object. This process has found industrial application various forms ranging from wood drying in the lumber industry to food drying in the food industry. In drying process, the heat may be added to the object from an external source by convection, conduction or radiation, or the heat can be generated internally within the solid body by means of electric resistance.

6.2 COMPUTER MODELS FOR HEAT FLOW

The effectiveness of a drying process depends on different factors: method of heat transfer, continuity or discontinuity of the process, direction of the heating fluids with respect to the product (pressure atmospheric, low, deep vacuum). Drying process can be performed by using different kinds of equipment such as: air cabinet, belt drier, tunnel drier, fluidized bed, spray drier, drum dryer, foam drier, freeze-drier, and microwave oven [1].

Radiation does not depend on any medium for its transmission. In fact, it takes place most freely when there is a perfect vacuum between the emitter and the receiver of such energy. This is proved daily by the transfer of energy from the sun to the earth across the intervening space. Radiation is a form of electromagnetic energy transmission and takes place between all matters providing that it is at a temperature above absolute zero. Infra-red radiation form just part of the overall electromagnetic spectrum. Radiation is energy emitted by the electrons vibrating in the molecules at the surface of a porous body. The amount of energy that can be transferred depends on the absolute temperature of the porous body and the radiant properties of the surface.

A porous body that has a surface that will absorb all the radiant energy it receives is an ideal radiator, termed a "black body". Such a porous body will not only absorb radiation at a maximum level but will also emit radiation at a maximum level. However, in practice, porous bodies do not have the surface characteristics of a black body and will always absorb, or emit, radiant energy at a lower level than a black body [2].

It is possible to define how much of the radiant energy will be absorbed, or emitted, by a particular surface by the use of a correction factor, known as the "emissivity" and given the symbol ε. The emissivity of a surface is the measure of the actual amount of radiant energy that can be absorbed, compared to a black body. Similarly, the emissivity defines the radiant energy emitted from a surface compared to a black body. A black body would, therefore, by definition, have an emissivity ε of 1.

Since World War II, there have been major developments in the use of microwaves for heating applications. After this time it was realized that microwaves had the potential to provide rapid energy-efficient heating of materials. These main applications of microwave heating today include food processing, wood drying, plastic, and rubber treating as well as curing and preheating of ceramics. Broadly speaking, microwave radiation is the term associated with any electromagnetic radiation in the microwave frequency range of 300 MHz–300 GHz. Domestic and industrial microwave ovens generally operate at a frequency of 2.45 GHz corresponding to a wavelength of 12.2 cm. However, not all materials can be heated rapidly by microwaves. Porous materials may be classified into three groups, that is conductors insulators, and absorbers. Porous materials that absorb microwave radiation are called dielectrics. Thus, microwave heating is also referred to as dielectric heating. Dielectrics have two important properties [3]:

They have very few charge carriers. When an external electric field is applied there is very little change carried through the material matrix.

The molecules or atoms comprising the dielectric exhibit a dipole movement distance. An example of this is the stereochemistry of covalent bonds in a water molecule, giving the water molecule a dipole movement. Water is the typical case of non-symmetric molecule. Dipoles may be a natural feature of the dielectric or they may be induced. Distortion of the electron cloud around non-polar molecules or atoms through the presence of an external electric field can induce a temporary dipole movement. This movement generates friction inside the dielectric and the energy is dissipated subsequently as heat.

The interaction of dielectric materials with electromagnetic radiation in the microwave range results in energy absorbance. The ability of a material to absorb energy while in a microwave cavity is related to the loss tangent of the material.

This depends on the relaxation times of the molecules in the material, which, in turn, depends on the nature of the functional groups and the volume of the molecule.

Generally, the dielectric properties of a material are related to temperature, moisture content, density, and material geometry.

An important characteristic of microwave heating is the phenomenon of "hot spot" formation, whereby regions of very high temperature form due to non-uniform heating. This thermal instability arises because of the non-linear dependence of the electromagnetic and thermal properties of material on temperature. The formation of standing waves within the microwave cavity results in some regions being exposed to higher energy than others. In this result, an increased rate of heating in these higher energy areas due to the non-linear dependence. Cavity design is an important factor in the control, or the utilization of this "hot spots" phenomenon [4].

Microwave energy is extremely efficient in the selective heating of materials as no energy is wasted in "bulk heating" the sample. This is a clear advantage that microwave heating has over conventional methods. Microwave heating processes are currently undergoing investigation for application in a number of fields where the advantages of microwave energy may lead to significant savings in energy consumption, process time, and environmental remediation.

Compared with conventional heating techniques, microwave heating has the following additional advantages:

(1) Higher heating rates.
(2) No direct contact between the heating source and the heated material.
(3) Selective heating may be achieved.
(4) Greater control of the heating or drying process.
(5) Reduced equipment size and waste.

As mentioned earlier, radiation is a term applied to many processes which involve energy transfer by electromagnetic wave (x-rays, light, and gamma rays). It obeys the same laws as light, travels in straight lines and can be transmitted through space and vacuum. It is an important mode of heat transfer encountered where large temperature difference occurs between two surfaces such as in furnaces, radiant driers, and baking ovens [5].

The thermal energy of the hot source is converted into the energy of electromagnetic waves. These waves travel through space into straight lines and strike a cold surface. The waves that strike the cold body are absorbed by that body and converted back to thermal energy or heat. When thermal radiations falls upon a body, part is absorbed by the body in the form of heat, part is reflected back into space and in some case part can be transmitted through the body.

The basic equation for heat transfer by radiation from a body at temperature T is:

$$q = A\varepsilon\sigma T^4$$

where ε is the emissivity of the body, $\varepsilon = 1$ for a perfect black body while real bodies which are gray bodies have an $\varepsilon < 1$

The porosity refers to volume fraction of void spaces. This void space can be actual space filled with air or space filled with both water and air. Many different definitions of porosity are possible. For non-hygroscopic materials, porosity does not change with change in moisture content. For hygroscopic materials, porosity changes with moisture content. However, such changes during processing are complex due to consideration of bound water and are typically not included in computations.

The distinction between porous and capillary-porous is based on the presence and size of the pores. Porous materials are sometimes defined as those having pore diameter greater than or equal to 10^{-7} m and capillary-porous as one having diameter less than 10^{-7} m. Porous and capillary porous materials were defined as those having a clearly recognizable pore space [6]. In non-hygroscopic materials, the pore space is filled with liquid if the material is completely saturated and with air if it is completely dry. The amount of physically bound water is negligible. Such a material does not shrink during heating. In non-hygroscopic materials, vapor pressure is a function of temperature only. Examples of non-hygroscopic capillary-porous materials are sand, polymer particles, and some ceramics. Transport materials in non-hygroscopic materials do not cause any additional complications as in hygroscopic materials.

In hygroscopic materials, there is large amount of physically bound water and the material often shrinks during heating. In hygroscopic materials there is a level of moisture saturation below which the internal vapor pressure is a function of saturation and temperature. These relationships are called equilibrium moisture isotherms. Above this moisture saturation, the vapor pressure is a function of temperature only and independent of the moisture level. Thus, above certain moisture level, all materials behave non-hygroscopic.

Transport of water in hygroscopic materials can be complex. The unbound water can be in funicular and pendular states. This bound water is removed by progressive vaporization under the surface of the solid, which is accompanied by diffusion of water vapor through the solid. Examples of porous materials are to be found in everyday life. Soil, porous or fissured rocks, ceramics, fibrous aggregates, sand filters, snow layers, and a piece of sugar or bread are but just a few. All of these materials have properties in common that intuitively lead us to classify them into a single denomination: porous media.

Indeed, one recognizes a common feature to all these examples. All are described as "solids" with "holes" that is presenting connected void spaces, distributed randomly or quite homogeneously within a solid matrix. Fluid flows can occur within the porous medium, so that we add one essential feature: this void space consists of a complex 3D network of interconnected small empty volumes called "pores", with several continuous paths linking up the porous matrix spatial extension, to enable flow across the sample [7].

If we consider a porous medium that is not consolidated, it is possible to derive the particle-size distribution of the constitutive solid grains. The problem is obvious when dealing with spherical shaped particles, but raises the question of what is meant by particle size in the case of an irregular shaped particle. In both cases, a first intuitive approach is to start with a sieve analysis. It consists to sort the constitutive solid particles among various sieves, each one having a calibrated mesh size. The most common type of sieve is a woven cloth of stainless steel or other metal, with wire diameter and tightness of weave controlled to produce roughly rectangular openings of known, uniform size. By shaking adequately the raw granular material, the solid grains are progressively falling through the stacked sieves of decreasing mesh sizes that is a sieve column. We finally get separation of the grains as function of their particle-size distribution that is also denoted by the porous medium granulometry. This method can be implemented for dry granular samples. The sieve analysis is a very simple and inexpensive separation method, but the reported granulometry depends very much on the shape of the particles and the duration of the laboratory test, since the sieve will let in theory pass any particle with a smallest cross-section smaller than the nominal mesh opening. For example, one gets very different figure while comparing long thin particles to spherical particles of the same weight.

The definition of a porous medium can be based on the objective of describing flow in porous media. A porous medium is a heterogeneous system consisting of a rigid and stationary solid matrix and fluid filled voids. The solid matrix or phase is always continuous and fully connected. A phase is considered a homogeneous portion of a system, which is separated from other such portions by a definitive boundary, called an interface. The size of the voids or pores is large enough such that the contained fluids can be treated as a continuum. On the other hand, they are small enough that the interface between different fluids is not significantly affected by gravity.

The topology of the solid phase determines if the porous medium is permeable, that is if fluid can flow through it, and the geometry determines the resistance to flown and therefore the permeability. The most important influence of the geometry on the permeability is through the interfacial or surface area between the solid phase and the fluid phase. The topology and geometry also determine if a porous medium is isotropic, that is all parameters are independent of orientation or anisotropic if the parameters depend on orientation. In multi-phase flow the geometry and surface characteristics of the solid phase determine the fluid distribution in the pores, as does the interaction between the fluids. A porous medium is homogeneous if its average properties are independent of location, and heterogeneous if they depend on location. An example of a porous medium is sand. Sand is an unconsolidated porous medium and the grains have predominantly point contact. Because of the irregular and angular nature of sand grains, many wedge-like crevices are present. An important quantitative aspect is the surface area of the sand grains exposed to the fluid. It determines the amount of water which can be held by capillary forces against the action of gravity and influences the degree of permeability [8].

The fluid phase occupying the voids can be heterogeneous in itself, consisting of any number of miscible or immiscible fluids. If a specific fluid phase is connected, continuous flow is possible. If the specific fluid phase is not connected, it can still have bulk movement in ganglia or drops. For single-phase flow the movement of a Newtonian fluid is described. For two-phase immiscible flow, a viscous Newtonian wetting liquid together with a non-viscous gas are described. In practice these would be water and air.

A detailed description of the complex 3D network of pores is obviously impossible to derive. For consolidated porous media, the determination of a pore size distribution is nevertheless useful. For those particular media, it is indeed impossible to handle any particle size distribution analysis.

One approach to define a pore size is in the following way: the pore diameter δ at a given point within the pore space is the diameter of the largest sphere that contains this point, while still remaining entirely within the pore space. To each point of the pore space such a "diameter" can be attached rigorously, and the pore size distribution can be derived by introducing the pore size density function $\theta(\delta)$ defined as the fraction of the total void space that has a pore diameter comprised between δ and $\delta + d\delta$. This distribution is normalized by the relation:

$$\int_0^{\infty} \theta(\delta)d\delta = 1$$

A porous structure should be:

    (1) A material medium made of heterogeneous or multiphase matter. At least one of the considered phases is not solid. The solid phase is usually called the solid matrix. The space within the porous medium domain that is not part of the solid matrix is named void space or pore space. It is filled by gaseous and/or liquid phases.

    (2) The solid phase should be distributed throughout the porous medium to draw a network of pores, whose characteristic size can vary greatly. Some of the pores comprising the void space must enable the flow across the solid matrix, so that they should then be interconnected [9].

    (3) The interconnected pore space is often denoted as the effective pore space, while unconnected pores may be considered from the hydrodynamic point of view as part of the solid matrix, since those pores are ineffective as far as flow through the porous medium is concerned. They are dead-end pores or blind pores, that contain stagnant fluid and no flow occurs through them.

A porous material is a set of pores embedded in a matrix of mostly solid material. The pores are the voids in the material itself. Pores can be isolated or interconnected. Furthermore, a pore can contain a fluid or a vapor, but it can also be empty. If the pore is completely filled with the fluid, it will be called saturated and if it is partially filled, it will be called non-saturated. So the porous material is primarily characterized by the content of its voids and not by the properties of the material itself.

Microwaves with their ability to rapidly heat materials are commonly used as a source of heat. In recent years, microwave drying has gained popularity as an alternative drying method in the food industry. The food industry is the largest consumer of microwave energy, where it can be employed for cooking, thawing, tempering, drying, freeze-drying, and sterilization, baking, heating, and re-heating. Microwave drying is rapid, more uniform and energy efficient compared to conventional hot air drying. Other advantages of microwave drying include space savings and energy efficiency, since most of the electromagnetic energy is converted into heat. Another advantage of microwave application for drying is the internal heat generation. In microwave processing the energy is transferred directly to the sample producing a volumetric heating [10].

There have been several experimental and theoretical studies on the analysis of heat and moisture transfer during drying of food products and on the determination of mass transfer characteristics such as moisture diffusion and mass transfer coefficient, undertaken by several researchers and engineers. The objective of any drying process is to produce a dried product of desired quality at minimum cost and maximum throughput possible. Microwave drying could be rapid, more uniform and energy efficient compared to conventional hot air drying.

Researchers investigated the effects of different drying methods on the color of the obtained products. They found that color characteristics are significantly affected by the drying methods. Scientists presented a review of reported experimental moisture diffusivity data in food materials.

Researchers developed new analytical models in a simple and accurate manner to determine the mass transfer characteristics for the geometrically shaped products. They also introduced new drying parameters in terms of drying coefficient and lag factors. Scientists presented a simple model of moisture transfer for multidimensional products. By considering the analogy between the heat diffusion and moisture transfer, drying time for infinite slab products was formulated. The analysis then extended to multidimensional products through the geometric shape factors introduced.

Scientists determined the effective moisture diffusivity of garlic cloves during a microwave-convective drying process. They also investigate its dependence on factors such as microwave power, air temperature, and air velocity that essentially influences drying rates [11].

Researchers determined the mass transfer characteristics for potato slab and cylinders subjected to convection, microwave and microwave-convective drying by adopting the analytical model proposed by Dincer and Dost. They have shown that the model is an effective means by which to calculate the mass transfer characteristics, also the result show that the power of the microwave has the main effect in drying.

In the present work, experimental data from a microwave drying system are used to determine the mass transfer characteristics for slab potato samples by adopting the analytical model developed by Dincer and Dost. Also a prediction model was presented by using factorial technique method for investigate the effect of microwave power and sample's dimensions on the drying characteristics. The model was applied successfully in the case of potato. The result shown the microwave power has the main effect and increase the dimensions of sample increase the drying time.

6.3 EXPERIMENTAL

The drying system used in this work was a microwave oven (Butan, model no. MF 45) of variable power output settings and rated capacity of 900 W at 2.45 GHz, outside dimensions (W x D x H), 601 x 465 x 338 mm and cavity dimensions (W x D x H), 419 x 428 x 245 mm. a schematic diagram microwave dryer is shown in Figure 1.

Trials were performed on potato tubers (Tarom potato) purchased at a local market in Rasht, Iran. This selection was based on several factors, including the fact that the potato cells are small, homogeneous, and structurally less complex than other vegetables. It was noted that composition of potato tubers depends upon generic and climatic factors. This may lead to some variations in the moisture content of potatoes within and between varieties. All potato tubers were washed in lukewarm water, hand-peeled, and cut into required dimensions [12].

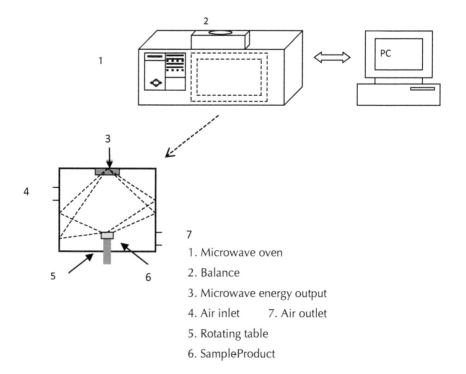

FIGURE 1   A schematic diagram of microwave drying equipment.

In all experiments, the microwave oven was brought to the operating temperature by heating 1,000 ml of distilled water in a glass beaker for 5 min before the first run of the day. The potato samples were placed on Petri dishes in the center of the microwave oven cavity. Throughout the experimental run the sample weights were continuously recorded at predetermined time intervals until no discernible difference between subsequent readings was observed. The moisture content value was determined as:

$$M = (W_t - W_d)/W_d \qquad (1)$$

where M is moisture content, W_t is the weight of sample (g) at any time and W_d is the weight of the dried sample.

6.4   ANALYSIS OF HEAT AND MOISTURE TRANSFER

A complete drying profile consists of two stages: a constant—rate period and a falling—rate period. It is frequently agreed that the mechanism of moisture movement within a hygroscopic solid during the falling-rate period could be represented by diffusion phenomenon according to Fick's second low. The governing Fickian equation is exactly in the form of the Fourier equation of heat transfer, in which temperature and thermal diffusivity are replaced with concentration and moisture diffusivity, respectively. Therefore, similar to the case of unsteady heat transfer, one can consider three different situations for the unsteady moisture diffusion, namely, the cases where the Biot number has the following values: $Bi \leq 0.1$, $0.1 < Bi < 100$, and $Bi > 100$. The first case, corresponding to situations where $Bi \leq 0.1$, imply negligible internal resistance to the moisture diffusivity within the solid object. On the other hand, cases where Bi, including negligible surface resistance to the moisture transfer at the solid object, are the most common situation, while cases where $0.1 < Bi < 100$, including the finite internal and surface resistances to the moisture transfer, exist in practical applications [13].

The time dependent heat and moisture transfer equations in Cartesian, cylindrical, and spherical coordinates for an infinite slab, infinite cylinder, and a sphere, respectively, can be written in the following compact form:

$$\left(\frac{1}{y^m}\right)\left(\frac{\partial}{\partial y}\right)\left[y^m\left(\frac{\partial T}{\partial y}\right)\right] = \left(\frac{1}{\alpha}\right)\left(\frac{\partial T}{\partial t}\right) \tag{2}$$

for heat transfer
and

$$\left(\frac{1}{y^m}\right)\left(\frac{\partial}{\partial y}\right)\left[y^m\left(\frac{\partial M}{\partial y}\right)\right] = \left(\frac{1}{D}\right)\left(\frac{\partial M}{\partial t}\right) \tag{3}$$

for moisture transfer.
where m = 0, 1, and 2 for an infinite slab, infinite cylinder, and a sphere. y = z for an infinite slab, y = r for infinite cylinder and sphere. T represents temperature (°C), M is moisture content by weight as dry basis (kg/kg), α is thermal diffusivity (m²/s), D is moisture diffusivity (m²/s), and t is time (s).

The dimensionless temperature (θ) and dimensionless moisture content (ϕ) can be defined as follows:

$$\theta = (T - T_i)/(T_a - T_i) \tag{4}$$

$$\phi = (M - M_e)/(M_i - M_e) \tag{5}$$

where subscripts a, e, and i indicate ambient, equilibrium, and initial conditions, respectively.

6.5 MODELING

Using the dimensionless moisture content (ϕ), the unsteady state diffusion of moisture in a food system by Fick's second low for an infinite slab can be expressed as:

$$\frac{\partial \phi}{\partial t} = \frac{\partial}{\partial z}(\frac{1}{D}\frac{\partial \phi}{\partial z}) \tag{6}$$

In order to simplify and solve this partial differential equation, the following hypotheses are made:
(1) The initial moisture content is uniform throughout the solid.
(2) The shape of the solid remains constant and shrinkage is negligible.
(3) The effect of heat transfer on mass transfer is negligible
(4) Mass transfer is by diffusion only.
(5) The moisture diffusion occurs in the z direction (perpendicular to the slab surface) only.

Under these assumptions, the governing 1D moisture diffusion equation, Equation (6), can be written as:

$$\begin{array}{lll} -D(\partial\phi(Y,t)/\partial z) = h_m\phi(Y,t) & \textit{for} & 1 \le Bi \le 100 \\ \phi(Y,t) = 0 & \textit{for} & Bi > 100 \end{array} \tag{7}$$

The following initial and boundary conditions are considered:

$$\phi(z,0) = 1 \tag{8}$$

$$(\partial\phi(0,t)/\partial z) = 0 \tag{9}$$

$$\begin{array}{lll} -D(\partial\phi(Y,t)/\partial z) = h_m\phi(Y,t) & \textit{for} & 1 \le Bi \le 100 \\ \phi(Y,t) = 0 & \textit{for} & Bi > 100 \end{array} \tag{10}$$

where Y is half thickness of slab and the Biot number is $Bi = h_m Y/D$.

Dincer and Dost developed a compact form of the equations for 1D transient moisture diffusion in an infinite slab. By applying the appropriate initial and boundary conditions, the governing equations were solved and further simplified to give the dimensionless moisture content at any point of the product in the following form:

$$\phi = \sum_{n=1}^{\infty} A_n B_n \tag{11}$$

The solution Equation (11) can be simplified if the values of $(\mu_1^2 Fo) > 1.2$ are negligibly small. Thus, the infinite sum in Equation (11) is well approximated by the first term only.

$$\phi \cong A_1 B_1 \tag{12}$$

where A_1 and B_1 are given by:

$$A_1 = \exp[(0.2533Bi)/(1.3 + Bi)] \tag{13}$$

$$B_1 = \exp\left(-\mu_1^2 F_0\right) \text{ for } Bi > 0.1. \tag{14}$$

where Fourier number is defined as $F_0 = Dt/Y^2$, Biot number is $Bi = h_m Y/D$, and Y is the characteristic dimension (half-thickness for slab).

Due to the fact that drying has an exponentially decreasing trend, the analysis assumed an exponential form for the dimensionless moisture distribution by introducing a lag factor (G, dimensionless) and drying coefficient (k, 1/s):

$$\phi = G \exp\left(-kt\right) \tag{15}$$

Drying coefficient shows the drying capability of an object or product per unit time and lag factor is an indication of internal resistance of an object to the heat and/or moisture transfer during drying. These parameters are useful in evaluating and representing a drying process. Both Equations (12) and (15) are in the same form and can be equated to each other and present a model for the moisture diffusivity:

$$D = \frac{kY^2}{\mu_1^2} \tag{16}$$

The coefficient, μ_1 for each object was determined by evaluating the root of the corresponding characteristic equation. For the purpose of practical drying applications, simplified expressions for the roots of the characteristic equations (μ_1) were developed as:

$$\mu_1 = \tan^{-1}(0.640443Bi + 0.380397) \tag{17}$$

The procedure used in evaluating and determining the process parameters is clearly given in Figure 2.

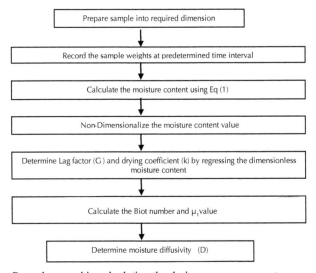

FIGURE 2   Procedure used in calculating the drying process parameters.

6.5.1 Factorial Technique Method

The two levels selected for each of the three variables are shown in Table 1. For the convenience of recording and processing the experimental data, the upper and lower levels of the variables were coded as +1 and –1, respectively and the coded values of any intermediate levels were calculated by using the expression:

$$X_i = \frac{X - \left(\dfrac{X_{max} + X_{min}}{2} \right)}{\left(\dfrac{X_{max} - X_{min}}{2} \right)} \tag{18}$$

where X_i is required coded value of a variable, X is any value of the variable from X_{min} to X_{max}, X_{min} is the lower level of the variable and X_{max} is the upper level of the variable.

TABLE 1   Controlling parameters.

			Level		Coding	
Parameter	**Notation**	**Unit**	**Low**	**High**	**Low**	**High**
Microwave power	P	W	90	450	–1	+1
Sample diameter	D	mm	20	40	–1	+1
Sample thickness	T	mm	3	10	–1	+1

Table 2 shows the 8 sets of coded conditions used to form the design matrix of 2^3 factorial design.

Some features of this table are:
(a) Trials indicate the sequence number of run under consideration,
(b) X_0 represents the mean parameter of the experiment,
(c) X_1, X_2 and X_3 represent the notation used for controlled variables in the order of microwave power, sample diameter and sample thickness, respectively, and
(d) The signs + 1 and – 1 as mentioned before refer to the upper and lower levels of that parameter under which they are recorded [14].

A mathematical function, f, was assumed to describe the relationship between drying constant k and the independent variables, such as, $k = f(P, D, T)$. According to experimental data which are shown graphically, the drying constant was assumed to vary linearly with each independent variable in the related interval. Hence, a first-order polynomial with interactions can be considered as the model, namely,

$k = b_0 + b_1 P + b_2 D + b_3 T + b_4 PD + b_5 PT + b_6 DT + b_7 PDT$ (19)

where k is the drying constant.

TABLE 2 Drying constants for potato samples as per design matrix.

Trial		P	D	T
Number	X0	X1	X2	X3
1	+1	−1	−1	−1
2	+1	+1	−1	−1
3	+1	−1	+1	−1
4	+1	+1	+1	−1
5	+1	−1	−1	+1
6	+1	+1	−1	+1
7	+1	−1	+1	+1
8	+1	+1	+1	+1

TABLE 3 Design matrix.

Trial Number	k1	k2
	0.1293	0.1223
	0.3921	0.3782
	0.1394	0.1475
	0.4297	0.4429
	0.1185	0.1088
	0.4224	0.4175
	0.1203	0.1288
	0.4575	0.4708

The main and interaction effects (ej) and coefficients (bj) were determined by using the formula:

$$e_j = 2b_j = \frac{2\sum_{i=1}^{N} X_{ij} k_i}{N} \tag{20}$$

where X_{ij} is the value of factor or interaction in the coded form, k_i is drying constant and N is the total number of observations.

The analysis of variance (ANOVA) technique was used to check the adequacy of the developed model. As per this technique, (a) The F-ratio of the developed model is calculated and is compared with the standard tabulated value of F-ratio for a specific level of confidence, (b) If the calculated value of F-ratio does not exceed the tabulated value, then with the corresponding confidence probability the model may be considered to be adequate. For this purpose the F-ratio of the model is defined as the ratio of variance of adequacy, also known as residual variance (usually denoted as S_{ad}^2) to the variance of reproducibility, also known as variance of optimization parameter (usually denoted as S_y^2). Therefore,

$$F_{\text{mod } el} = \frac{S_{ad}^2}{S_y^2} \tag{21}$$

Here,

$$S_{ad}^2 = \sum_{i=1}^{N} \frac{(k_i - \hat{k}_i)^2}{DF} \tag{22}$$

where N is the number of trials, k_i is observed (or measured from experiments) response, $\hat{k}$ is predicted/estimated value of the response (i.e., the one obtained from the model), DF is degrees of freedom and it is equal to [N–(K+1)] where K represents the number of independently controllable variables and

$$S_y^2 = \frac{\sum\limits_{q=1}^{2} \sum\limits_{i=1}^{N} (k_{iq} - \bar{k}_i)^2}{N} \tag{23}$$

where k_{iq} is the value of response in a repetition, q is the number of repetition and $\bar{k}_i$ is the arithmetical mean of repetitions (i.e., response in the repetitions).

To recognize the significant coefficients, the student's t-test is used. As per this test, (1) The calculated value of t corresponding to a coefficient is compared with the standard tabulated value of specific level of probability, (2) If the calculated value of t exceeds the tabulated one, then with the corresponding confidence probability the coefficient is said to be significant. For this purpose the value of t is given by:

$$t = \frac{b_j}{S_{bj}} \tag{24}$$

where $\left| b_j \right|$ represent the absolute value of coefficient whose significance is being tested and S_{bj} the standard deviation of coefficients given by:

$$S_{bj}^2 = \frac{\mathrm{var}\ iance\ of\ optimizati\ on}{No\ of\ trials} = \frac{S_y^2}{N} \tag{25}$$

S_{bj}, alternatively, called as variance of the regression coefficients, is thus seen to be same for all the coefficients. Thus, they depend only on the error of the experiments and the confidence interval.

6.6 DISCUSSION AND RESULTS

Throughout the experimental run the sample weights were continuously recorded at regular time intervals until no discernible difference between subsequent readings was observed. Then the moisture ratio of the samples was determined from Equation (5). A typical drying curve for potato slab is shown in Figure 3.

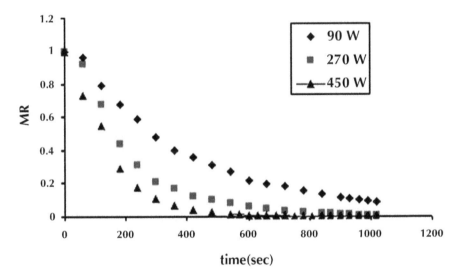

FIGURE 3 Drying curves of potato slab (diameter: 40 mm).

The dimensionless moisture content values (calculated by Equation (5)) were then regressed against the drying time in the exponential form of Equation (15) using the least square curve fitting method. Thus, the drying coefficients (k) and lag factors (G) were determined for samples as presented in Tables 4. Using the calculated lag factor, the Bi number for each experimental condition was determined using Equations (13), as appropriate. Subsequently, the associated values of μ_1 were computed from the simplified expression for a slab (Equation (17)). The calculated Biot numbers and μ_1 values are shown in Table 5.

The drying coefficient (k) is a parameter which indicates the drying capability of the solid object. The effect of microwave power on the drying constant is shown Table 4. The ability of microwaves to facilitate rapid drying rates was observed in magnitude of the coefficients, which increased with increasing output power level; for example slab (40 mm radius) 0.0024, 0.0048, and 0.0077 s^{-1} for 90, 270, and 450 W respectively. As expected, during microwave drying, the variable power had the most significant effect on the drying capability.

An increase in slab diameter results an increase in the drying coefficient (Table 4). It is because of sudden and volumetric heating, generating high pressure inside the potato samples, resulted in boiling and bubbling of the samples. On the other hand, the amount of water in the sample increases, without increasing the resistance of it, and results in faster drying. At low microwave power (90 W), the drying coefficient increases slightly with increase in sample diameter. However, at higher power (450 W), the drying coefficient grows at a higher rate. As mentioned before, this is because of higher water content, and hence, more absorption of microwave power, results in faster drying of samples. Using the values of Y, k, and μ_1, the moisture diffusivity (D) was then computed from Equation (16). The calculated diffusivity values are shown in Table 6.

The final mathematical model as determined by this method is in the form of

$$k = 0.2766 + 0.1498 \, P + 0.0155 \, D + 0.0084 \, PD + 0.0117 \, PT$$

The developed model has been found to be adequate by ANOVA technique as shown in Table 7. This model shows that the drying rate increases with increasing the microwave power or sample diameter. As mentioned the effect of microwave power is so higher than sample diameter.

The significant interaction effects between variables are shown in Figure.4 and Figure 5. It is seen from Figure 4 that at low microwave power, about 90 W, the drying constant increases only slightly with an increase in sample diameter. However, at higher microwave power, the drying constant increases at a higher rate with an increase in sample diameter. The interaction effect between microwave power and sample thickness is shown in Figure 5. The effect of thickness at low power level on the drying coefficient is more considerable. It is observed that the drying constant increases with an increase in microwave power which is obviously expected. However, below the microwave power of 270 W the drying constants for thicker samples are numerically lower than those for thinner plates. This could possibly due to increasing internal resistance to mass transfer. However, beyond a microwave power of 270 W, the trend is reversed. This is because that in high level, the effect of microwave power is more than the effect of increasing internal resistance to mass transfer.

TABLE 4   Drying coefficient and lag factor values for microwave drying of potato slabs.

Experimental conditions			k (s–1)	G
Microwave power (W)	diameter (mm)	thickness (mm)		
90	20	3	0.0021	1.056

TABLE 4 *(Continued)*

Experimental conditions			k (s–1)	G
Microwave power (W)	diameter (mm)	thickness (mm)		
90	40	3	0.0024	1.057
270	20	3	0.0042	1.082
270	40	3	0.0048	1.079
450	20	3	0.0064	1.086
450	40	3	0.0077	1.089

TABLE 5 Mass transfer characteristics for microwave drying of potato slabs.

Experimental conditions			μ_1	*Bi*
Microwave power (W)	Diameter (mm)	Thickness (mm)		
90	20	3	0.625	0.51
90	40	3	0.605	0.491
270	20	3	0.653	0.61
270	40	3	0.568	0.637
450	20	3	0.64	0.668
450	40	3	0.658	0.683

TABLE 6 Moisture diffusivity values for microwave drying of potato slabs.

Experimental conditions			$D \times 10^{-8}$ (m^{2}s^{-1})
Microwave power (W)	Diameter (mm)	Thickness (mm)	
90	20	3	1.25
90	40	3	1.46
270	20	3	2.24
270	40	3	2.62
450	20	3	3.23
450	40	3	3.76

TABLE 7   Analysis of variance (ANOVA).

Param-eter	Degree of Freedom	Variance of Opti-mization Param-eter	Standard Deviation of Coef-ficients,	Variance of Ad-equacy	'F'-ratio (Model) at (4,8,0.05)	'F'- ratio Model from Tables where Adequate		
S_y^2	S_{ad}^2	S_y^2	S_{bj}	S_{ad}	F_m	F_t	F_m	$< F_t$
k	8	4	5.31e–5	0.0026	5.31e–5	1	3.84	yes

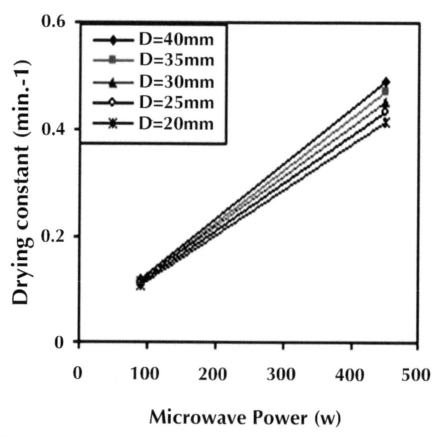

FIGURE 4.   Effect of parameter interaction between P and D (at T = 10 mm).

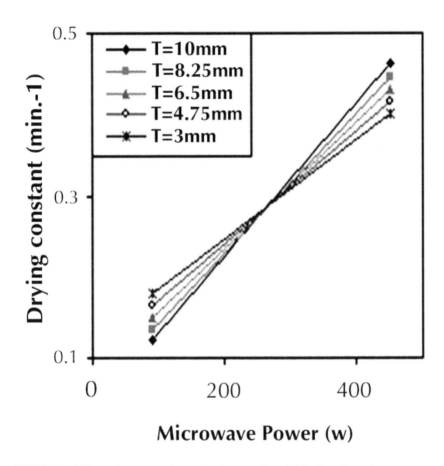

FIGURE 5 Effect of parameter interaction between P and T (at D = 40 mm).

6.7 CONCLUSION

Based on the results of this study, the following conclusions were drawn:

(1) Drying took place mainly in the falling rate period followed by a constant rate period after a short heating period.

(2) The drying rate increases with increasing the microwave power or sample diameter.

(3) An increase in slab diameter results an increase in the drying cocfficient. It is because of sudden and volumetric heating, generating high pressure inside the potato samples, resulted in boiling and bubbling of the samples.

(4) At low microwave power (90 W), the drying coefficient increases slightly with increase in sample diameter.

(5) The variable power had most significant effect on the drying capability.

(6) Drying constant increases with an increase in microwave power which is obviously expected.

(7) Below the microwave power of 270 W the drying constants for thicker samples are numerically lower than those for thinner plates. This could possibly due to increasing internal resistance to mass transfer. However, beyond a microwave power of 270 W, the trend is reversed. This is because that in high level, the effect of microwave power is more than the effect of increasing internal resistance to mass transfer.

(8) In order to maximize the benefits of microwave drying, further studies are required at lower power outputs with different microwave power cycles.

(9) In further studies, more comprehensive experimental application of the method should be considered to attain a better understanding drying process of potato slabs as a function of time. Moreover, the influence of various sizes of potato slabs can be studied using this approach.

KEYWORDS

- **Analysis of variance**
- **Black body**
- **Dielectrics**
- **Emissivity**
- **Factorial technique method**

REFERENCES

1. Armour, J. and Cannon, J. Fluid Flow through Woven Screens. *AIChE J.*, **14**(3), 415–420 (1968).
2. ASTM D737-75, Standard Test Methods for Air Permeability of Textile Fabrics.
3. ASTM E96-95, Water Vapor Transmission of Materials.
4. Arnold, G. and Fohr, J. P. Slow Drying Simulation in Thick Layers of Granular Products. *Int. J. Heat Mass Transfer*, **31**(12), 2517–2562 (1988).
5. Azizi, S., Moyne, C., and Degiovanni, A. Approches Experimental theorique de la Conductivite Thermique des Milieux Poreux Humides. *Int. J. Heat Mass Transfer*, **31**(11), 2305–2317 (1988).
6. Backer, S. The Relationship between the Structural Geometry of a Textile Fabric and its Physical Properties, Part IV: Interstice Geometry and Air Permeability. *Textile Res. J.*, **21**, 703–714 (1951).
7. Barnes, J., and Holcombe, B. Moisture Sorption and Transport in Clothing during Wear. *Textile Res. J.*, **66**(12), 777–786 (1996).
8. Bartles, V. T. *Survey on the Moisture Transport Properties of Foul Weather Protective Textiles at Temperatures around and Below the Freezing Point.* Technical Report no. 11674, Hohenstein Institute of Clothing Physiology, Boennigheim Germany (2001).

9. Bears, J. *Dynamics of Fluids in Porous Media*. Elsevier, New York (1972).

10. Black, W. Z. and Hartley, J. G. *Thermodynamics*. Harper and Row, New York (1985).

11. BS 4407, *Quantitative Analysis of Fiber Mixtures* (1997).

12. BS 7209, *Specification for Water Vapor Permeable Apparel Fabrics* (1990).

13. CAN2-4.2-M77, Method of Test for Resistance of Materials to Water Vapor Diffusion (Control Dish Method) (1977).

7 Heat Flow and Porous Materials

CONTENTS

7.1 INTRODUCTION

For heat flow analysis of wet porous materials, the liquid is water and the gas is air. Evaporation or condensation occurs at the interface between the water and air so that the air is mixed with water vapor. A flow of the mixture of air and vapor may be caused by external forces, for instance, by an imposed pressure difference. The vapor will also move relative to the gas by diffusion from regions where the partial pressure of the vapor is higher to those where it is lower.

7.2 COMPUTER MODELS FOR HEAT FLOW IN WET POROUS MATERIALS

The heat flow in porous media is the study of energy movement in the form of heat which occurs in many types of processes. The transfer of heat in porous media occurs from the high to the low temperature regions. Therefore, temperature gradient has to exist between the two regions for heat transfer to happen. It can be done by conduction (within one porous solid or between two porous solids in contact), by convection (between two fluids or a fluid and a porous solid in direct contact with the fluid), by radiation (transmission by electromagnetic waves through space) or by combination of the three methods.

The general equation for heat transfer in porous media is:

$$\begin{pmatrix} rate\ of \\ heat\ in \end{pmatrix} + \begin{pmatrix} rate\ of\ generation \\ of\ heat \end{pmatrix} = \begin{pmatrix} rate\ of \\ heat\ out \end{pmatrix} + \begin{pmatrix} rate\ of\ accumulation \\ of\ heat \end{pmatrix}$$

When a wet porous material is subjected to thermal drying two processes occur simultaneously, namely:

(1) Transfer of heat to raise the wet porous media temperature and to evaporate the moisture content (MC).
(2) Transfer of mass in the form of internal moisture to the surface of the porous material and its subsequent evaporation.

The rate at which drying is accomplished is governed by the rate at which these two processes proceed. Heat is a form of energy that can across the boundary of a system. Heat can, therefore be defined as "the form of energy that is transferred between a system and its surroundings as a result of a temperature difference". There can only be a transfer of energy across the boundary in the form of heat if there is a temperature difference between the system and its surroundings. Conversely, if the system and surroundings are at the same temperature there is no heat transfer across the boundary [1].

Strictly speaking, the term "heat" is a name given to the particular form of energy crossing the boundary. However, heat is more usually referred to in thermodynamics through the term "heat transfer" which is consistent with the ability of heat to raise or lower the energy within a system.

There are three modes of heat flow in porous media:

(1) Convection
(2) Conduction
(3) Radiation

All three are different. Convection relies on movement of a fluid in porous material. Conduction relies on transfer of energy between molecules within a porous solid or fluid. Radiation is a form of electromagnetic energy transmission and is independent of any substance between the emitter and receiver of such energy. However, all three modes of heat flow rely on a temperature difference for the transfer of energy to take place.

The greater the temperature difference the more rapidly will the heat be transferred. Conversely, lower the temperature difference: the slower will be the rate at which heat is transferred. When discussing the modes of heat transfer it is the rate of heat transfer Q that defines the characteristics rather than the quantity of heat.

As it was mentioned earlier, there are three modes of heat flow in porous structures, convection, conduction, and radiation. Although two, or even all three, modes of heat flow may be combined in any particular thermodynamic situation, the three are quite different and will be introduced separately.

The coupled heat and liquid moisture transport of porous material has wide industrial applications. Heat transfer mechanisms in porous textiles include conduction by the solid material of fibers, conduction by intervening air, radiation, and convection. Meanwhile, liquid and moisture transfer mechanisms include vapor diffusion in the void space and moisture sorption by the fiber, evaporation, and capillary effects. Water vapor moves through porous textiles as a result of water vapor concentration differences. Fibers absorb water vapor due to their internal chemical compositions and structures. The flow of liquid moisture through the textiles is caused by fiber liquid molecular attraction at the surface of fiber materials, which is determined mainly by surface tension and effective capillary pore distribution and pathways. Evaporation and/or condensation take place, depending on the temperature and moisture distributions. The heat transfer process is coupled with the moisture transfer processes with phase changes such as moisture sorption/desorption and evaporation/condensation [2].

All three of the mechanisms by which heat is transferred conduction, radiation, and convection, may enter into drying. The relative importance of the mechanisms varies from one drying process to another and very often one mode of heat transfer predominates to such extent that it governs the overall process.

As an example, in air drying the rate of heat transfer is given by:

$$q = h_s A \left(T_a - T_s \right)$$

where q is the heat transfer rate in Js^{-1}, h_s is the surface heat transfer coefficient in Jm^{-2}s^{-1} °C^{-1}, A is the area through which heat flow is taking place, m^{-2}, T_a is the air temperature and T_s is the temperature of the surface which is drying, °C.

To take another example, in a cylindrical dryer where moist material is spread over the surface of a heated cylinder, heat transfer occurs by conduction from the cylinder to the porous media, so that the equation is:

$$q = UA \left(T_i - T_s \right)$$

where U is the overall heat transfer coefficient, T_i is the cylinder temperature (usually very close to that of the steam), T_s is the surface temperature of textile and A is the area of the drying surface on the cylinder. The value of U can be estimated from the conductivity of the cylinder material and of the layer of porous solid [3].

Mass transfer in the drying of a wet porous material will depend on two mechanisms: movement of moisture within the porous material which will be a function of the internal physical nature of the solid and its MC, and the movement of water vapor from the material surface as a result of water vapor from the material surface as a result of external conditions of temperature, air humidity and flow, area of exposed surface and supernatant pressure.

Some porous materials such as textiles exposed to a hot air stream may be cooled evaporative by bleeding water through its surface. Water vapor may condense out of damp air onto cool surfaces. Heat will flow through an air-water mixture in these situations, but water vapor will diffuse or convect through air as well. This sort of transport of one substance relative to another called mass transfer. The MC, X, is described as the ratio of the amount of water in the materials, m_{H2O} to the dry weight of material $m_{material}$:

$$X = \frac{m_{H2O}}{m_{material}}$$

There are large differences in quality between different porous materials depending on structure and type of material. A porous material such as textiles can be hydrophilic or hydrophobic. The hydrophilic fibers can absorb water, while hydrophobic fibers do not. A textile that transports water through its porous structures without absorbing moisture is preferable to use as a first layer. Mass transfer during drying depends on the transport within the fiber and from the textile surface, as well as on how the textile absorbs water, all of which will affect the drying process [4-6].

As the critical MC or the falling drying rate period is reached, the drying rate is less affected by external factors such as air velocity. Instead, the internal factors due to moisture transport in the material will have a larger impact. Moisture is transported in porous media during drying through:

(1) Capillary flow of unbound water
(2) Movement of bound water
(3) Vapor transfer

Unbound water in a porous media will be transported primarily by capillary flow. As water is transported out of the porous material, air will be replacing the water in the pores. This will leave isolated areas of moisture where the capillary flow continues.

Moisture in a porous structure can be transferred in liquid and gaseous phases. Several modes of moisture transport can be distinguished:

(1) Transport by liquid diffusion
(2) Transport by vapor diffusion
(3) Transport by effusion (Knudsen-type diffusion)
(4) Transport by thermo-diffusion
(5) Transport by capillary forces
(6) Transport by osmotic pressure
(7) Transport due to pressure gradient.

A very common method of removing water from porous structures is convective drying. Convection is a mode of heat transfer that takes place as a result of motion within a fluid. If the fluid starts at a constant temperature and the surface is suddenly increased in temperature to above that of the fluid, there will be convective heat transfer from the surface to the fluid as a result of the temperature difference. Under these conditions the temperature difference causing the heat transfer can be defined as:

= surface temperature – mean fluid temperature

Using this definition of the temperature difference, the rate of heat transfer due to convection can be evaluated using Newton's law of cooling:

$$Q = h_c A \Delta T$$

where A is the heat transfer surface area and is the coefficient of heat transfer from the surface to the fluid, referred to as the "convective heat transfer coefficient".

The units of the convective heat transfer coefficient can be determined from the units of other variables:

$$Q = h_c A \Delta T$$
$$W = (h_c) m^2 K$$

so the units of are .

The relationships given in Equations (0.4 and 0.5) are also true for the situation where a surface is being heated due to the fluid having higher temperature than the surface [7-9]. However, in this case the direction of heat transfer is from the fluid to the surface and the temperature difference will now be

mean fluid temperature – surface temperature

The relative temperatures of the surface and fluid determine the direction of heat transfer and the rate at which heat transfer take place.

As given in equations, the rate of heat transfer is not only determined by the temperature difference but also by the convective heat transfer coefficient. This is not a constant but varies quite widely depending on the properties of the fluid and the behavior of the flow. The value of must depend on the thermal capacity of the fluid particle considered, that is for the particle. So, higher the density and of the fluid is better convective heat transfer.

Two common heat transfer fluids are air and water, due to their widespread availability. Water is approximately 800 times denser than air and also has a higher value. If the argument given is valid then water has a higher thermal capacity than air and should have a better convective heat transfer performance. This is borne out in practice because typical values of convective heat transfer coefficients are as follows:

Fluid	
water	500–10,000
Air	5–100

The variation in the values reflects the variation in the behavior of the flow, particularly the flow velocity, with the higher values of resulting from higher flow velocities over the surface. When a fluid is in forced or natural convective motion along a surface, the rate of heat transfer between the solid and the fluid is expressed by the following equation:

$$q = h.A\left(T_W - T_f\right)$$

The coefficient h is dependent on the system geometry, the fluid properties, velocity, and the temperature gradient. Most of the resistance to heat transfer happens in the stationary layer of fluid present at the surface of the solid, therefore the coefficient h is often called film coefficient [10-12].

Correlations for predicting film coefficient h are semiempirical and use dimensionless numbers which describe the physical properties of the fluid, the type of flow, the temperature difference, and the geometry of the system.

The Reynolds number characterizes the flow properties (laminar or turbulent). L is the characteristic length: length for a plate, diameter for cylinder or sphere.

$$N_{Re} = \frac{\rho L \nu}{\mu}$$

The Prandtl number characterizes the physical properties of the fluid for the viscous layer near the wall.

$$N_{Pr} = \frac{\mu c_p}{k}$$

The Nusselt number relates the heat transfer coefficient h to the thermal conductivity k of the fluid.

$$N_{Nu} = \frac{hL}{k}$$

The Grashof number characterizes the physical properties of the fluid for natural convection.

$$N_{Gr} = \frac{L^3 \Delta \rho g}{\rho \gamma^2} = \frac{L^3 \rho^2 g \beta \Delta T}{\mu^2}$$

In capillary porous materials, moisture migrates through the body as a result of capillary forces and gradients of MC, temperature, and pressure. This movement contributes to other heat transfer mechanisms while eventual phase change occurring within the material act as heat sources or sinks.

Drying is fundamentally a problem of simultaneous heat and mass transfer under transient conditions resulting in a system of coupled non-linear partial differential equations. Scientists defined a coupled system of partial differential equations for heat and mass transfer in porous bodies. Although, they used different approaches to obtain equations, their formulations do not differ substantially from each other. Many numerical works have been executed in this field, on basis of these two theories.

Researchers have developed a 1D model for simultaneous heat and moisture transfer in porous materials. Also many scientists have used 1D in studying heat and mass transfer during convective drying of porous media [13-17].

Researchers studied drying problem of timber, with a 2D model. Many scientists used finite element method for solution of 2D heat and mass transfer in porous media.

All listed studies, have estimated heat and mass transfer between porous materials and drying fluid by coefficients obtained from standard correlations based on boundary layer equations and more of them assumed analogy between heat and mass transfer coefficients. However, since the actual process of drying is a conjugate problem, the heat and mass transfer, to and from the porous solid have to be studied along with the flow field.

Scientists in a conjugate study of paper drying have shown that results of solution by conjugate view differ considerably from those of decoupled system. Also the analogy between heat and mass transfer coefficient may not exist in reality, even in drying of 1D objects.

Researchers found that the mentioned analogy holds good only for initial period of unsaturated sand drying, and for the later part of drying, the heat and mass transfer coefficients at the interface may does not satisfy the analogy, due to the non-uniformity of moisture and temperature distribution at the interface resulting from conjugate nature of transfers [18, 19].

Scientists studied drying of wood as a conjugate problem. They have used boundary layer equations for flow field, and presented temperature and moisture contours during the process.

Researchers applied 2D model for brick drying. They used Navier-Stokes equations for flow field including buoyancy terms in their conjugate analysis. They concluded that restricting heat and mass transfer to top surface of 2D porous body will cause considerable errors into solution. They also have shown that neglecting buoyant forces in flow analysis, leads to considerable differences in heat and mass transfer values and lower drying rate, in Reynolds number of 200.

In the majority of the studies, the buoyancy effects have been simply ignored in flow field analysis except the study performed by some researchers. The solution method used is finite volume approach and is related to an unsteady problem. In this chapter we have tried to use a much conservative method for calculation of energy and momentum fluxes. Regarding the weak capability of the finite element methods (specially in their flux averaging steps), the motivation of the current work concerns on solving the same 2D conjugate problem with a finite volume approach which fundamentally guaranties the energy and the mass fluxes to be conserved during the solution and the descritization procedures. However, using the same mesh and consequently the same cells, entire the solution domain highlights the ability of finite volume approach on using unstructured grids. On the other hand, the solution is extended to a higher Reynolds numbers to give a wider range study of the buoyancy effects not only on temperature and the MC, but also on flow patterns during the drying process.

7.3 MODELING

The problem model, including corresponding boundary conditions is shown in Figure 1. The problem considers a sample of rectangular brick exposed to convective airflow. The brick is assumed to be saturated with water initially. The governing moisture removal from brick to air exceeds in cause of concentration gradient between air in vicinity of body and free stream air. Flow is incompressible and thermophysical properties are taken to be constant. The initial MC of the brick is 0.13 kg/kg of dry solid, and the solid temperature is set to 293 K and the drying air has a 50% of relative humidity. At the Reynolds 200 the air velocity is 0.02 m/s. The thermal conductivity of the brick is 1.8 W/mK, and the thermal capacity is set to 1,200 J/kgK. The brick density is set to 1,800 kg/m³.

Non-isothermal diffusion coefficient of porous body in vapor phase, $D_{tv} = 1(10)^{-12}$

Non-isothermal diffusion coefficient of porous body in liquid phase, $D_{tl} = 1(10)^{-12}$

Iso-thermal diffusion coefficient of porous body in vapor phase, $D_{mv} = 1(10)^{-12}$

Iso-thermal diffusion coefficient of porous body in liquid phase, $D_{ml} = 1(10)^{-8}$

Enthalpy of evaporation (Initial value), $H_{fg} = 2454 Kj / KgK$

Mass diffusion coefficient of vapor in air, $Diff - 0.256(10)^{-4}$

Fluid thermal conductivity, $K_f = 0.02568$

Porous body thermal conductivity, $K_s = 1.8$

Bouyancy coefficient of temperature, $\beta = 3.4129(10)^{-3} 1/K$

Bouyancy coefficient of concentration, $\beta' = 0.0173$

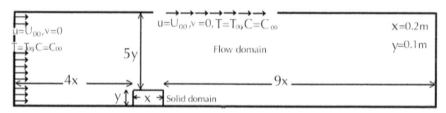

FIGURE 1   Geometry of computational field.

7.3.1 Governing Equations for Solid (Brick)

The equations for porous solid phase as obtained by researchers on the basis of continuum approach, were applied for numerical solution:

- Energy equation:

$$c^* \frac{\partial T}{\partial t} = \left(\frac{k}{\rho_0} + h_{fg} D_{tv} \right) \left(\frac{\partial^2 T}{\partial x^2} + \frac{\partial^2 T}{\partial y^2} \right) + h_{fg} D_{mv} \left(\frac{\partial^2 M}{\partial x^2} + \frac{\partial^2 M}{\partial y^2} \right) \tag{1}$$

where

$$c^* = c_0 + m_l c_l + m_v c_v$$

Moisture conservation equation:

$$\frac{\partial M}{\partial t} = (D_{tl} + D_{tv}) \left(\frac{\partial^2 T}{\partial x^2} + \frac{\partial^2 T}{\partial y^2} \right) + (D_{ml} + D_{mv}) \left(\frac{\partial^2 M}{\partial x^2} + \frac{\partial^2 M}{\partial y^2} \right) \tag{2}$$

7.3.2 Governing Equations for Flow Field

- Continuity

$$\frac{\partial u}{\partial x} + \frac{\partial v}{\partial y} = 0 \tag{3}$$

- Momentum Equation (2D Navier-Stokes):

$$\frac{\partial u}{\partial t} + u\frac{\partial u}{\partial x} + v\frac{\partial u}{\partial y} = -\frac{1}{\rho}\frac{\partial P}{\partial x} + \upsilon\left(\frac{\partial^2 u}{\partial x^2} + \frac{\partial^2 u}{\partial y^2}\right) \tag{4}$$

$$\frac{\partial v}{\partial t} + u\frac{\partial v}{\partial x} + v\frac{\partial v}{\partial y} = -\frac{1}{\rho}\frac{\partial P}{\partial y} + \upsilon\left(\frac{\partial^2 v}{\partial x^2} + \frac{\partial^2 v}{\partial y^2}\right) + g\beta(T - T_\infty) + g\beta'(C - C_\infty) \tag{5}$$

- Energy equation:

$$\frac{\partial T}{\partial t} + u\frac{\partial T}{\partial x} + v\frac{\partial T}{\partial y} = \alpha\left(\frac{\partial^2 T}{\partial x^2} + \frac{\partial^2 T}{\partial y^2}\right) \tag{6}$$

- Vapor concentration equation:

$$\frac{\partial C}{\partial t} + u\frac{\partial C}{\partial x} + v\frac{\partial C}{\partial y} = D\left(\frac{\partial^2 C}{\partial x^2} + \frac{\partial^2 C}{\partial y^2}\right) \tag{7}$$

7.3.3 Boundary and Initial Conditions

Initially the porous material is assumed to be at uniform MC (saturation value), and temperature (equal to air temperature).

$$T(x,y,0) = T_0, \quad M(x,y,0) = M_0$$

The boundary condition at interface of solid and fluid are:

- No slip condition

$$u = 0; \quad v = 0$$

- Continuity of temperature

$$T_f = T_s$$

- Continuity of concentration

$$C = C(T,M)_s$$

- Heat balance

$$\left(k + \rho_0 h_{fg} D_{tv}\right)\frac{\partial T}{\partial n} + \rho_0 h_{fg} D_{mv}\frac{\partial M}{\partial n} = k_f\frac{\partial T_f}{\partial n} + h_{fg} D\frac{\partial C}{\partial n} \tag{8}$$

- Species flux balance

$$\rho_0 \left(D_{tv} \frac{\partial T}{\partial n} + D_{mv} \frac{\partial M}{\partial n} \right) = D \frac{\partial C}{\partial n} \tag{9}$$

7.3.4   Boundary Conditions for Flow Field

- Inlet boundary condition

$$u = U_\infty, \; v = 0, T = T_\infty, C = C_\infty$$

- Far stream boundary condition (upper boundary)

$$u = U_\infty, \; T = T_\infty, C = C_\infty$$

- Outflow boundary condition

$$\frac{\partial u}{\partial x} = 0$$

Also bottom surface of solid is taken adiabatic.

7.4   GRID DEPENDENCY

A structured mesh was used for computational work. Mesh clustering around body is illustrated in Figure 2.

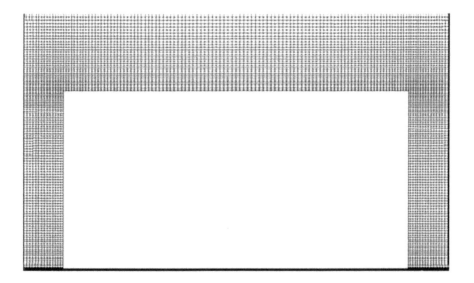

FIGURE 2   Mesh structure over the porous material.

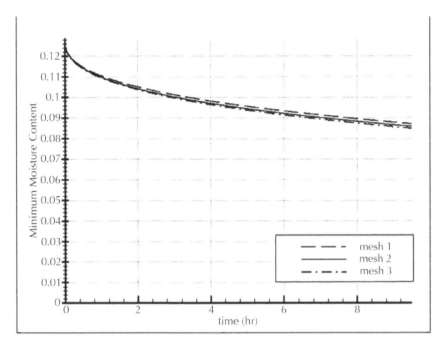

FIGURE 3   Value of MC in leading edge obtained with three different meshes.

To clarify effect of mesh refinement on numerical solution, three meshes with different precision were used in numerical analysis:

Mesh 1: with 69120 total nodes.
Mesh 2: with 108000 total nodes.
Mesh 3; with 155520 total nodes.

As shown in Figure 3 value of MC in leading edge (minimum value of moisture content) obtained by numerical solution with Mesh 1 and Mesh 2 in 9 hr have maximum difference of 1.28%, while for Mesh 2 and Mesh 3 the maximum difference is 0.66%. So Mesh 2 seems to be optimum in accuracy and run-time, and therefore was decided to continue the computational work [20].

7.5 NUMERICAL SOLUTION

In each time step, following items should be carried out:

(1) Solving 2D flow field equations (continuity + NS) by simple algorithm, with finite volume scheme.
(2) Solution of energy equation for fluid by ADI technique with finite difference scheme. Neumann boundary condition was used for interface, which obtained from last time step derivative value of solid temperature in Equation (8).
(3) Then energy equation for porous field is solved by ADI technique with finite difference scheme. Boundary condition at interface is the known value of fluid temperature.

(4) Determining concentration distribution for the flow field. The solution is similar with step 2, but boundary condition is obtained from Equation (9) by explicit MC derivative at interface.

(5) Calculation of MC distribution for porous body. The solution is similar with
step 3, but boundary value of moisture is obtained by known fluid concentration (from previous step) value at interface.

Because of using explicit values in the solution procedure, we have to do internal
repetition in each time step, until internal conversion is reached for all four variables.
After that, solution for next time step starts.

7.6 DISCUSSION AND RESULTS

7.6.1 Validation and General Results

Drying curve of modeled brick in Re = 200 is verified here. As illustrated in Figure 4,
this comparison reveals that samples possessed almost the nearly same trend of MC
reduction in the overall drying time of 15 hr. Nevertheless, the accuracy of mentioned
curves are within 0.6%.

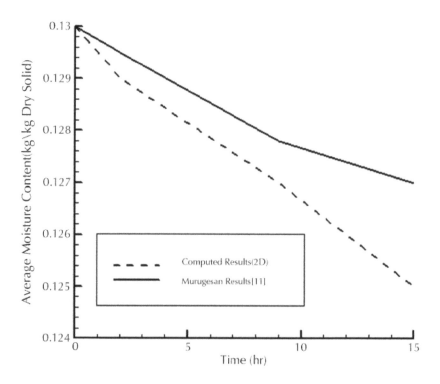

FIGURE 4   Drying curve.

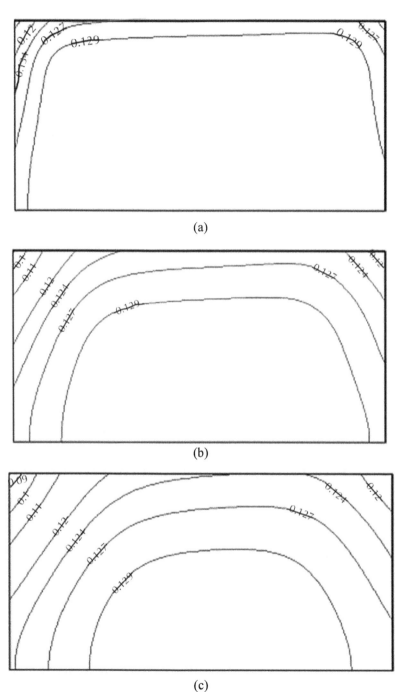

(a)

(b)

(c)

FIGURE 5   Moisture content distributions (kg moisture/kg dry) during drying. (a) t = 21/2 hr, (b) t = 9 hr, and (c) t = 151/2 hr.

The computational results of MC profile for Re = 200 illustrates that drying rate in region near leading edge (which corresponds to maximum concentration gradient in adjacent air) is more than other regions in porous body (Figure 5). Gradually, drying spreads from that region to centric regions of body.

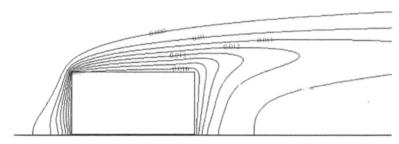

FIGURE 6   Concentration contours in air around porous body.

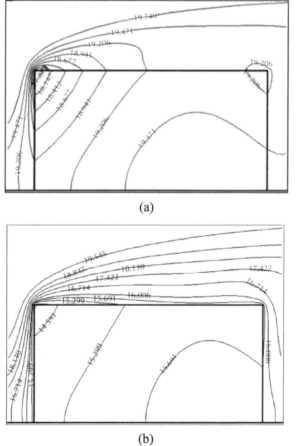

(a)

(b)

FIGURE 7   Temperature distributions in porous body and around. (a) t = 21/2 hr and (b) t = 151/2 hr.

Concentration values in air around porous body are presented in Figure 6. Concluding from this figure, gradient of concentration in the body surface has shown strong effect on MC distribution in body (as seen for leading edge).

Temperature distributions in porous body and around air in the course of drying are shown in Figure 7. It is evident that, temperature value of porous body near leading edge decreases quickly as a result of higher moisture vaporization from surface there and this temperature drop transfer to centric regions of porous body.

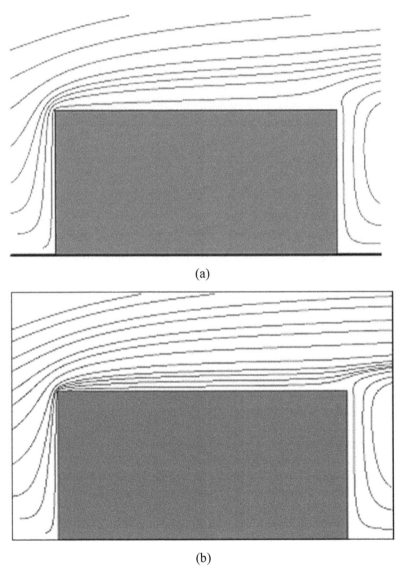

(a)

(b)

Figure 8.   *(Continued)*

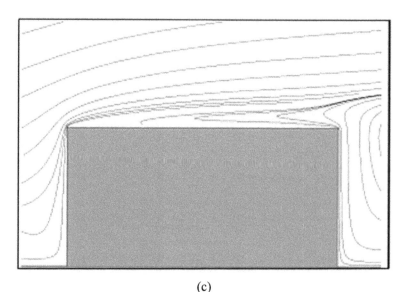

(c)

FIGURE 8 Streamlines for different value of Re. (a) Rey = 50, (b) Rey = 100, and (c) Rey = 200.

7.6.2 Different Air Velocities

Numerical solution was executed for different velocities of drying fluid (different Reynolds number) to clarify the effect of this parameter. Figure 8(a)–(c) shows streamlines around body for Reynolds number of 50, 100, and 200. As shown in this figure, velocity increment in Re = 50 to Re = 100, results in more compactness of streamlines above body, and then in Re = 200 separation occurs, while a vortex forms on upper surface.

In Figure 9 contours of concentration around body for denoted numbers of Reynolds are illustrated. It is obvious that Reynolds increment results a significant decrement in thickness of concentration boundary layer on the upper side (especially for leading edge) due to changes in streamlines and velocity boundary layer.

For Re = 100, MCs are less than those obtained for Re = 50, especially for regions nearby leading edge (Figure 10(a) and (b)). This is mainly due to thinner concentration boundary layer aforementioned. In transition to Re = 200 from Re = 100 (Figure 10(b) and (c)), this matter satisfies just for leading edge and left side of body (as a result of vortex formation above body in Re = 200).

In Figure 11 drying curves are shown for various Reynolds number ranging from 50 to 1,000. Effect of drying fluid velocity on process speed could be clearly analyzed. For example, removed moisture after 5 hr of drying process for Re =100 is 15.4% more than corresponding value of Re = 50. This difference is 17.5% for two Reynolds numbers of 500 and 200.

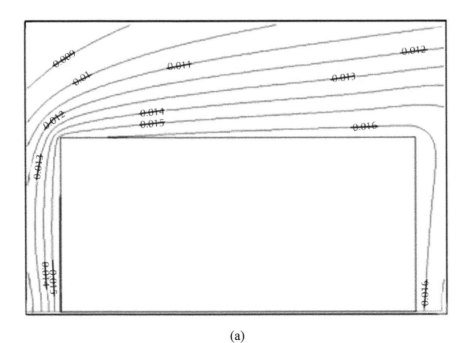

(a)

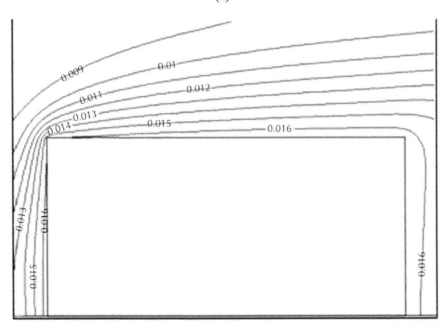

(b)

FIGURE 9   *(Continued)*

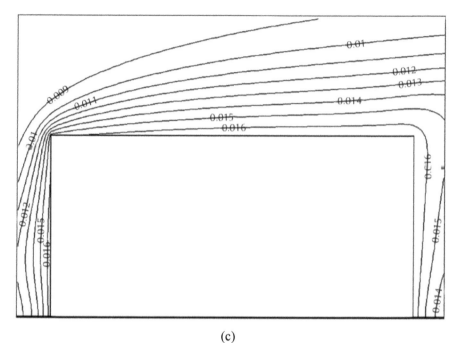

(c)

FIGURE 9   Concentration contours for different value of Re. (a) Rey = 50, (b) Rey = 100, and (c) Rey = 200.

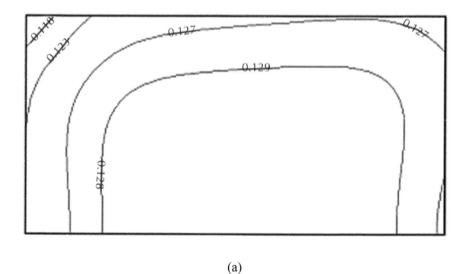

(a)

FIGURE 10   *(Continued)*

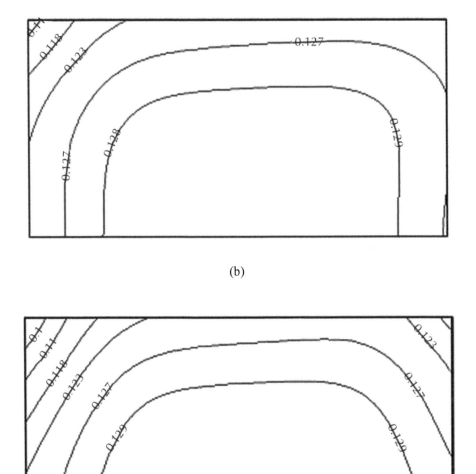

(b)

(c)

FIGURE 10   Moisture content contours (kg moisture /kg dry) for different value of Re. (a) Rey = 50, (b) Rey=100, and (c) Rey = 200.

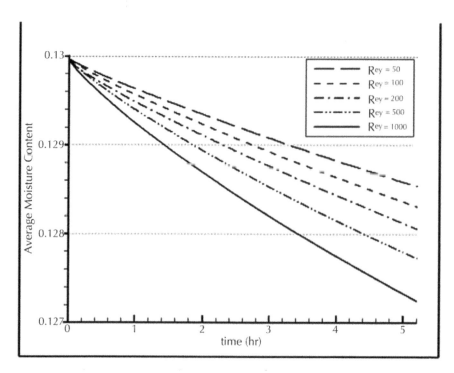

FIGURE 11   Drying curves for Re = 50 to 100.

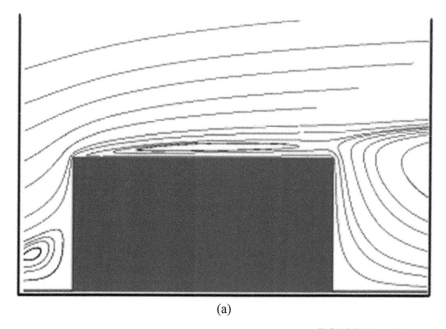

(a)

FIGURE 12   *(Continued)*

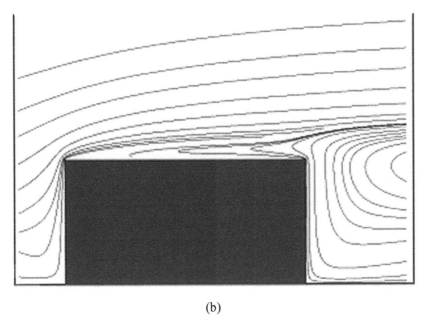

(b)

FIGURE 12   Effect of buoyancy forces on streamlines in Re = 200. (a) Forced convection, and (b) Mixed convection.

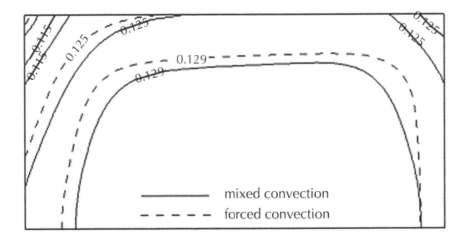

FIGURE 13   Moisture profiles obtained by mixed and forced convection models in Re = 200 (t = 8 1/4 hr).

7.6.3 The Effect of Buoyancy

To study the contribution of buoyancy on flow patterns and consequently on drying process, the computations were performed with and without buoyancy terms in flow equations (i.e. mixed and forced convection, respectively). Figure 12 shows streamlines around porous body in Re = 200 for both of forced and mixed convection assumptions. As shown, buoyancy forces clustered the streamlines near vertical walls. In Figure 13 the MC distributions was shown for $8\frac{1}{4}$ duration for both cases (Re = 200). Obviously, the mixed convection results show a higher drying performance.

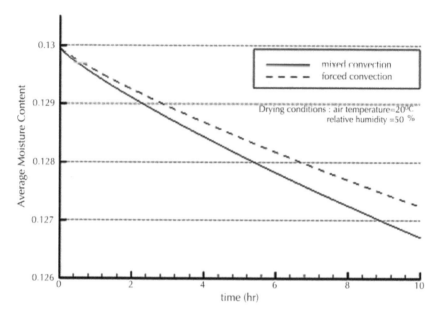

FIGURE 14   Drying curve for mixed and forced convection models in Re = 200.

The comparison between drying curves of mixed and forced assumptions are shown in Figure 14. The decrease of MC has a lower rate in forced convection case for 10 hr period, (e.g. 19.7% after 10 hr of drying). Consequently, the forced convection model underestimates drying rate noticeably in Re = 200.

The buoyancy forces has a great contribution in drying prediction in Re = 200. To investigate effect intensity in various drying fluid velocities (different Reynolds numbers), drying curves resulted by mixed and forced convection models in Reynolds numbers 50 and 1000 (a practical range in drying) are illustrated in Figure 15. Also Figure 16 shows average moisture fluxes (during initial 5 hr of process) obtained by each of two models, and so in Table 1 percentage increase in average moisture flux by taking buoyancy into account, are listed for different Reynolds numbers. These figures and table implies that despite relative decreasing in drying rate with increasing Reynolds, the effect of buoyancy on drying process in whole of Reynolds range of 50–1,000, are considerable. So, in the denoted range of Re, which include most practical velocities in porous bodies drying (especially clay products drying), ignoring buoyancy effects in flow analysis, will impose noticeable error into computations.

In other words, forced convection model has not enough accuracy for drying process analysis in governing range [21-23].

TABLE 1   Effect of buoyancy on average moisture flux for different Reynolds numbers.

Re	percentage increase in average moisture flux
50	26%
100	20%
200	19%
500	16%
1,000	15%

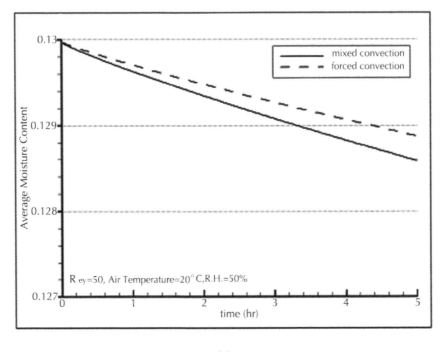

(a)

FIGURE 15   *(Continued)*

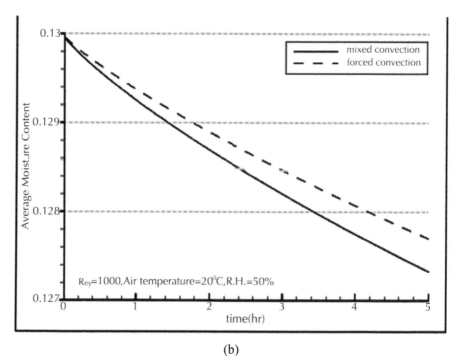

(b)

FIGURE 15 Drying curve for mixed and forced convection models Re. (a) Re = 50, and (b) Re = 1,000.

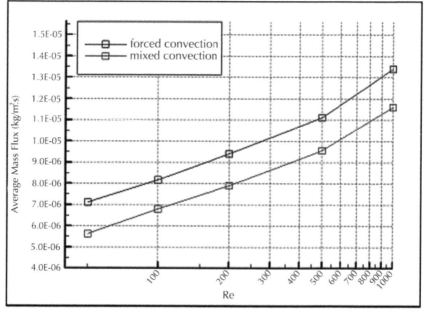

FIGURE 16 Average mass flux during initial 5 hr of drying for mixed and forced convection models.

7.7 CONCLUSION

Using a FV method which employed in the pressure based conservative algorithms, performed a 2D conjugated solution that guaranties the conservation laws despite of the finite elements methods.

By studying drying process in a variety of flow velocity (various Reynolds number), It is observed that airflow velocity increment has proportional effect on drying rate, with a factor between $1/_4$ and $1/_5$.

Moisture profiles and drying rate are considerably affected by buoyancy forces. Moisture removal from porous body surface for mixed convection model is more than forced convection in whole of Reynolds range of 50–1,000. So, it is suggested to taking into consideration buoyancy effects in analysis of flow around porous bodies in drying or other similar processes.

KEYWORDS

- **Buoyancy forces**
- **Diffusion coefficient**
- **Navier–Stokes equations**
- **Porous material**
- **Prandtl number**
- **Reynolds number**

REFERENCES

1. Incropera, F. P. and Dewitt, D. P. *Fundamentals of Heat and Mass Transfer*, second ed., Wiley, New York (1985).
2. Ghali, K., Jones, B., and Tracy, E. Modeling Heat and Mass Transfer in Fabrics. *Int. J. Heat Mass Transfer*, **38**(1), 13–21 (1995).
3. Flory, P. J. *Statistical Mechanics of Chain Molecules*. Interscience Pub., New York (1969).
4. Hadley, G. R. Numerical Modeling of the Drying of Porous Materials. In *Proceedings of The Fourth International Drying Symposium, Vol. 1*, pp. 151–158 (1984).
5. .Hong, K., Hollies, N. R. S., and Spivak, S. M. Dynamic Moisture Vapor Transfer through Textiles, Part I: Clothing Hygrometry and the Influence of Fiber Type. *Textile Res. J.*, **58**(12), 697–706 (1988).
6. Chen, C. S. and Johnson, W. H. Kinetics of Moisture Movement in Hygroscopic Materials, In Theoretical Considerations of Drying Phenomenon. *Trans. ASAE.*, **12**, 109–113 (1969).
7. .Barnes, J. and Holcombe, B. Moisture Sorption and Transport in Clothing during Wear. *Textile Res. J.*, **66**(12), 777–786 (1996).
8. .Chen, P. and Pei, D. A Mathematical Model of Drying Process. *Int. J. Heat Mass Transfer*, **31**(12), 2517–2562 (1988).
9. Davis, A., and James, D. Slow Flow through a Model Fibrous Porous Mcdium. *Int. J. Multiphase Flow*, **22**, 969–989 (1996).
10. Jirsak, O., Gok, T., Ozipek, B., and Pau, N. Comparing Dynamic and Static Methods for Measuring Thermal Conductive Properties of Textiles. *Textile Res. J.*, **68**(1), 47–56 (1998).
11. Kaviany, M. *Principle of Heat Transfer in Porous Media*. Springer, New York (1991).

12. Jackson, J. and James, D. The Permeability of Fibrous Porous Media. *Can. J. Chem. Eng.*, **64**, 364–374 (1986).
13. Dietl, C., George, O. P., and Bansal, N. K. Modeling of Diffusion in Capillary Porous Materials during the Drying Process. *Drying Technol.*, **13**(1, 2), 267–293 (1995).
14. Ea, J. Y. *Water Vapor Transfer in Breathable Fabrics for Clothing*. PhD thesis, University of Leeds (1988).
15. Haghi, A. K. Moisture permeation of clothing. *JTAC*, **76**, 1035–1055 (2004).
16. Haghi, A. K. Thermal analysis of drying process. *JTAC*, **74**, 827–842 (2003).
17. Haghi, A. K. *Some Aspects of Microwave Drying*. The Annals of Stefan cel Mare University, Year VII, No. 14, 22–25 (2000).
18. Haghi, A. K. A *Thermal Imaging Technique for Measuring Transient Temperature Field: An Experimental Approach*. The Annals of Stefan cel Mare University, Year VI, No. 12, 73–76 (2000).
19. Haghi, A. K. Experimental Investigations on Drying of Porous Media using Infrared Radiation. *Acta Polytechnica*, **41**(1), 55–57 (2001).
20. Haghi, A. K., A Mathematical Model of the Drying Process. *Acta Polytechnica*, **41**(3), 20–23 (2001).
21. Haghi, A. K. Simultaneous Moisture and Heat Transfer in Porous System. *Journal of Computational and Applied Mechanics*, **2**(2), 195–204 (2001).
22. Haghi, A. K. A Detailed Study on Moisture Sorption of Hygroscopic Fiber. *Journal of Theoretical and Applied Mechanics*, **32**(2), 47–62 (2002).
23. Flory, P. J. *Statistical Mechanics of Chain Molecules*. Interscience Pub., New York (1969).

8 Thermal Environment

CONTENTS

8.1 INTRODUCTION

The thermal environment is sometimes very complex. Convection, radiation, and conduction are the common means of heat exchange and they vary independently over time and location. The final effects on the surface heat exchange of the human body are important factors for heat balance and for perception of the thermal conditions. Assessment of the thermal environment in modern office or car can create difficulties due to the complex interaction of the ventilation system with the situation close to the person and the external environmental factors (e.g. radiation, air temperature, and air movements). Furthermore, measurements in reality, as well as in the laboratory, contain various methodological problems.

8.2 COMPUTER MODELS FOR THERMAL ENVIRONMENT

Information on the transmission of water vapor by textiles fibers as porous materials is desirable for better understanding of the problems of comfort and data for design in special applications such as upholstery, footwear, immersion suits, and other protective clothing, and wrapping or packaging, where high resistance to liquid water is desired, combined with considerable permeability of water vapor. Nevertheless, some of the issues of clothing comfort that are most readily understood involve the mechanisms by which clothing materials influence heat and moisture transfer from the skin to the environment. It should be noted that heat transfer by convection, conduction, and radiation, moisture transfer by vapor diffusion are the most important mechanisms in very cool or warm environments [1].

During physical activity the body provides cooling partly by producing insensible perspiration. If the water vapor cannot escape to the surrounding atmosphere the relative humidity of the microclimate inside the clothing increases causing a corresponding increased thermal conductivity of the insulating air and the clothing becomes uncomfortable. In extreme cases hypothermia can result if the body loses heat more rapidly than it is able to produce it, for example, when physical activity has stopped, causing decrease in core temperature. If perspiration cannot evaporate and liquid sweat (sensible perspiration) is produced, the body is prevented from cooling at the same rate as heat is produced, for example, during physical activity and hyperthermia can result as the body core temperature increases. Table 1 shows heat energy produced by various activities and corresponding perspiration rates [2].

The ability of fabric to allow water vapor to penetrate is commonly known as breathability. This should more scientifically be referred to as water vapor permeability. Although perspiration rates and water vapor permeability are usually quoted in units of grams per day and grams per square meter per day respectively, the maximum work rate can only be endured for very short time. During rest, most surplus body heat is lost by conduction and radiation, whereas during physical activity, the dominant means of losing excess body heat is by evaporation of perspiration. It has been found that the length of time the body can endure arduous work decreases linearly with decrease in fabric water vapor permeability [3].

TABLE 1   Typical heat energy produced by various activities and corresponding perspiration rates

Activity	Work rate (watts)	Perspiration rate (g/day)
Sleeping	58–60	2,270–2,280
Sitting	95–100	3,750–3,800
Gentle Walking	185–200	7,550–7,600
Active Walking	285–300	11,480–11,500
With light pack	385–400	15,180–15,200
With heavy pack	487–500	18,900–19,000
Mountain walking with heavy pack	610–810	22,700–30,400
Maximum work rate	1,100–1,200	38,300–45,610

It has also been shown that the maximum performance of a subject wearing clothing with a vapor barrier is some 60% less than that of a subject wearing the same clothing but without a vapor barrier. Even with two sets of clothing that exhibit a small variation in water vapor permeability, the differences in the wearer's performance are significant.

In an environment where body temperature cannot be regulated without a lot of sweating, we often try to get rid of heat from our body by turning on the air conditioning systems or moving into a conditioned room. Just after the change of the environment, we will feel "cool" or "comfortable". But the sweat accumulated in clothing evaporates gradually, until the heat loss from our body can be more than needed and at last we might feel "cold" or "uncomfortable". A review of clothing studies has shown that moisture collection in cold weather clothing, even after heavy exercise, seldom exceeds 10% by weight of added water. One of the measurements is used to calculate values related to water vapor transmission properties are water "vapor resistance". This is the water vapor pressure difference across the two faces of the fabric divided by the heat flux per unit area, measured in square meters Pascal per watt. Some water vapor resistance (WVR) data on different types of outwear fabrics are presented in Table 2. The measurement of WVR in thickness unit (mm) is the thickness of a still air layer having the same resistance as the fabric. The use of thickness units facilitates the calculations of resistance values for clothing assemblies comprising textile and air layers.

This observation is generally explained by noting that the major transfer mechanism from wet skin to underwear is one of distillation. An initial observation noting the surprisingly strong discomfort sensations associated with small amounts of water in the skin clothing interface.

TABLE 2   Typical water vapor resistance (WVR) of fabrics.

Fabric, Outer (shell) material	WVR(mm still air)
Neoprene, rubber or PVC coated	1,000–1,200
Conventional PVC coated	300–400
Waxed cotton	1,000+
Wool overcoating	6–13
Leather	7–8
Woven microfiber	3–5
Closely woven cotton	2–4
Two-layer PTFE laminates	2–3
Three-layer laminates (PTFE, polyester)	3–6
Microporous polyurethane (various types)	3–14

It has been confirmed in a number of studies in which either moisture from sweating or added moisture generates these clothing contact sensations. The procedures for these measurements emphasize again that very little moisture is required to stimulate sensations of discomfort. Often 3–5% added moisture is ample to develop discomfort.

Simultaneous differential equations for the transfer of heat and moisture in porous media under combined influence of gravity and gradients of temperature and moisture content were developed by researchers. They have performed a general analysis of moisture migration in a slab of an unsaturated porous material for a condition where the temperature of one surface is suddenly increased to a higher value whereas the temperature of the other surface is maintained constant. Scientists has derived a general, 1D, steady-state model, describing the heat and mass transfer within a homogeneous porous medium, saturated with wetting liquid, its vapor and non-condensable gas. The effects of gas diffusion, phase change, conduction, liquid and vapor transport, capillarity, and gravity are included. The analysis is based on a general thermodynamic description of the unique equilibrium states characteristics of liquid wetting porous media. Researchers provided a systematic, rigorous, and unified treatment of the governing equations for simultaneous heat and mass transfer within a wide range of porous media.

Some work has also been done in the area of coupled diffusion of moisture and heat in hygroscopic textile materials. Scientists have given a review of numerical modeling of convection, diffusion, and phase changes in textiles. It summarizes current and past work aimed at utilizing CFD techniques for clothing applications. It was shown that water in a hygroscopic porous textile may exist in vapor or liquid form in the pore spaces. Phase changes associated with water include liquid evaporation/condensation in the pore spaces and sorption/desorption from polymer fibers. Additional factors such as swelling of solid polymer due to water and heat of sorption was incorporated into the appropriate conservation and transport equations. Many researchers attempted to solve the non-linear differential equations which describe coupled diffusion of heat and mass (moisture) in hygroscopic textile materials. In addition to the diffusion equations, a rate equation was introduced describing the rate of exchange of moisture between the solid (textile fibers) and the gas phase. The predictions compared favorably with experimental observations on wool bales and wool fabrics. Scientists developed a simple model of combined heat and water vapor transport in clothing. Transport by forced convection was not included in this model.

Researchers reported a strong dependency of WVR of hydrophilic membranes or coatings: higher the relative humidity at the membrane, lower the WVR (i.e., the higher the water vapor permeability or breathability).

In a temperature dependent experiment, scientists placed a hydrophilic film on an ice block. Water vapor sublimating from the ice could diffuse only through the film and was collected by a desiccant. Researchers measured mass transport through the film, and he found that WVR is an exponential function of temperature. In this experiment, water vapor permeability varnishes nearly completely with decreasing textile temperature. Because diffusion in hydrophilic materials is non-Fickian, he also derived from his results a theory of diffusion speed depending on activation energy, and he accounted for different relative humidity. Additionally, scientists reported an increase in the moisture vapor transmission rate of hydrophilic and microporous textiles when measuring with a heated dish instead of unheated dish. They interpreted their results by the increased motion of water vapor and polymer molecules, which they claimed would also work for microporous constructions [4].

Researcher compared cotton, water repellent cotton, and acrylic garments through wearing tests and concluded that the major factor causing discomfort was the excess amount of sweat remaining on the skin surface. Scientists stated that the ability of fabrics to absorb liquid water (sweat) is more important than water vapor permeability in determining the comfort factor of fabrics.

Researchers postulated physiological factors related to the wearing comfort of fabrics as follows: sweating occurs whenever there is a tendency for the body temperature to rise, such as high temperature in the surrounding air and physical exercise, and so on. If liquid water (sweat) cannot be dissipated quickly, the humidity of the air in the space in between the skin and the fabric that contacts with the skin rises. This increased humidity prevents rapid evaporation of liquid water on the skin and gives the body the sensation of "heat" that triggered the sweating in the first place. Consequently, the body responds with increased sweating to dissipate excess thermal energy. Thus a fabric's inability to remove liquid water seems to be the major factor causing uncomfortable feelings for the wearer [5].

Scientists conducted wearer trails for shirts made of various fibers and concluded that the largest factor that influenced wearing comfort was the ability of fibers to absorb water, regardless of wearer fibers were synthetic or natural.

All of these studies indicate that the transient state phenomenon responding to the physiological demand to cause sweating is most relevant to comfort or discomfort associated with fabrics.

When work is performed in heavy clothing, evaporation of sweat from the skin to the environment is limited by layers of wet clothing and air. The magnitude of decrement in evaporative cooling is a function of the clothing's resistance to permeation of water vapor. Researchers conducted an experimental study on the rate of absorption of water vapor by wool fibers. They observed that, if a porous textile is immersed in a humid atmosphere, the time required for the fibers come to equilibrium with this atmosphere is negligible compared with the time required for the dissipation of heat generated or absorbed when the regain changes. They investigated the effects of heat of sorption in the wool water sorption system. They observed that the equilibrium value of the water content was directly determined by the humidity but that the rate of absorption and desorption decreased as the heat-transfer efficiency decreased. Heat transfer was influenced by the mass of the sample, the packing density of the fiber assembly, and the geometry of the constituent fibers. Scientists pointed out that the water-vapor uptake rate of wool is reduced by a rise in temperature that is due to the heat of sorption. The dynamic water-vapor sorption behavior of fabrics in the transient state will therefore not be the same as that of single fibers owing to the heat of sorption and the process to dissipate the heat released or absorbed [6].

Researchers proposed a system of differential equations to describe the coupled heat and moisture diffusion into bales of cotton. Two of the equations involve the conservation of mass and energy, and the third relates fiber moisture content with the moisture in the adjacent air. Since these equations are non-linear, Henry made a number of simplifying assumptions to derive an analytical solution.

In order to model the two-stage sorption process of wool fibers, Scientists proposed three empirical expressions for a description of the dynamic relationship between fiber moisture content and the surrounding relative humidity. By incorporating several features omitted by scientists into the three equations. They were able to solve the model numerically. Since their sorption mechanisms (i.e. sorption kinetics) of fibers were neglected, the constants in their sorption-rate equations had to be determined by comparing theoretical predictions with experimental results. Based on conservation equations, this global model consists of two differential coupled equations with variables for temperature and water concentration in air and in the fibers of the textile, which is generally the water adsorbed by hygroscopic fibers. is not in equilibrium with , but an empirical relation between the adjustable parameters is assumed: the rate of sorption is a linear function of the difference between the actual and the equilibrium value. The introduced coefficients are not directly linked to the physical properties of the clothes [7].

Researchers reported a numerical model describing the combined heat and water-vapor transport through clothing. The assumptions in the model did not allow for the complexity of the moisture-sorption isotherm and the sorption kinetics of fibers. They presented two mechanical models to simulate the interaction between moisture sorption by fibers and moisture flux through the void spaces of a fabric. In the first model, diffusion within the fiber was considered to be so rapid that the fiber moisture content was always in equilibrium with the adjacent air. In the second model, the sorption kinetics of the fiber was assumed to follow Fickian diffusion. In these models, the effect of heat of sorption and the complicated sorption behavior of the fibers were neglected.

Scientists developed a two-stage model, which takes into account water-vapor sorption kinetics of wool fibers and can be used to describe the coupled heat and moisture transfer in wool fabrics. The predictions from the model showed good agreement with experimental observations obtained from a sorption-cell experiment. More recently, they further improved the method of mathematical simulation of the coupled diffusion of the moisture and heat in wool fabric by using a direct numerical solution of the moisture diffusion equation in the fibers with two sets of variable diffusion coefficients. These research publications were focused on fabrics made from one type of fiber. The features and differences in the physical mechanisms of coupled moisture and heat diffusion into fabrics made from different fibers have not been systematically investigated [8].

Scientists compared the heat exchange and thermal insulation of two ensembles, one made from wool, the other from nylon, worn by subjects who exercised either lightly (dry condition) or strenuously (wet condition) for 60 min, then rested 60 min. He found that there was a significant difference in physiological and subjective responses between dry and wet conditions, but not between the two fiber types. Further, there was no significant difference between the ratings of temperature and humidity sensations for the wool and nylon garments. The wool garment picked up more water than the nylon garment (245 g versus 198 g) for the wet condition.

However, the wool fabric may have been slightly thicker than the nylon fabric, since it was reported to have a slightly greater thermal resistance and therefore hold more water.

Researchers evaluated the effect of five kinds of knit structures: all made from 100% polypropylene were evaluated. On subjects exercising for 40 min at 5°C followed by 20 min at rest, and then repeated. The thickest knit, a fleece, caused the greatest total sweat production, retained the most moisture, and wetted skin the most. They stated that the hydrophobic polypropylene prevented extensive sweat accumulation in the underwear (10–22%) causing the sweat to accumulate in the outer garments [9].

Scientists repeated the protocol above, but used low and high work rates with three kinds of underwear (a polypropylene 1 1 knit, a wool 1 1 knit, and a fishnet polypropylene) worn under wool fleece covered by polyester/cotton outer garments. Total sweat production and evaporated sweat were the same for all three underwear fabrics, but where the sweat accumulated differed significantly. More sweat accumulated in the wool underwear than either polypropylene at both work rates. At the higher work rate, more sweat moved into the fleece layer from both kinds of polypropylene underwear than for the wool. Most likely for the 1 1 knits, the thicker wool underwear(1.95 mm) simply holds more water than the polypropylene underwear (1.41 mm) and based on outer layer to layer wicking results, needs a greater volume of sweat to fill it pores before it starts to donate the excess to the layer above it.

Researchers conducted wear trails for shirts made of various fibers. They concluded that the largest factor that influenced wearing comfort was the ability of fibers to absorb water, regardless of whether fibers were synthetic or natural.

All of these studies indicate that the transient state phenomenon responding to the physiological demand to cause sweating is most relevant to comfort or discomfort associated with this general principle. It is important to point out that a highly water absorbing fabric placed in the first layer keeps the partial pressure of water vapor near the skin low, which helps dissipate water at the skin surface, although the water vapor transport rate is smaller than for non-absorbing fabrics.

In the other words, the dissipation of water by means of absorption by fabrics appears to be much more efficient way to keep the water vapor pressure near the skin low than dissipation by permeation through fabrics. Highly water absorbing fabrics raise the temperature of the air space near the skin. The temperature rise will further decrease relative humidity: however, the higher temperature may or may not desirable depending on environmental conditions [10].

In the literature, the emphasis has been placed on the correlation between sweating and discomfort associated with wearing fabrics. However, there is relatively less emphasis placed on the influence of changes in the surrounding conditions, that is, the influence of the seasons. Many comfort studies are conducted with a single layer of fabric at relatively warm and moderately humid conditions. Severe winter conditions, which mandate the use of layered fabrics, would necessitate totally different kinds of testing procedures. Consequently, it is necessary to distinguish the comfort factor and the survival factor, and to investigate these factors with different perspective.

The evaporation process is also influenced by the liquid transport process. When liquid water cannot diffuse into the fabric, it can only evaporate at the lower surface of the fabric. As the liquid diffuses into the fabric due to capillary action, evaporation can take place throughout the fabric.

Moreover, the heat transfer process has significant impact on the evaporation process in cotton fabrics but not in polyester fabrics. The process of moisture sorption is largely affected by water vapor diffusion and liquid water diffusion, but not by heat transfer. When there is liquid diffusion in the fabric, the moisture sorption of fibers is mainly determined by the liquid transport process, because the fiber surfaces are covered by liquid water quickly. Meanwhile, the water content distributions in the fibers are not significantly related to temperature distributions [11].

All moisture transport processes, on the other hand, affect heat transfer significantly. Evaporation and moisture sorption have a direct impact on heat transfer, which in turn is influenced by water vapor diffusion and liquid diffusion. The temperature rise during the transient period is caused by the balance of heat released during fiber moisture sorption and the heat absorbed during the evaporation process.

As a whole, a dry fabric exhibits three stages of transport behavior in responding to external humidity transients. The first stage is dominated by two fast processes: water vapor diffusion and liquid water diffusion in the air filling the interfiber void spaces, which can reach new steady states within fractions of seconds. During this period, water vapor diffuses into the fabric due to the concentration gradient across the two surfaces. Meanwhile, liquid water starts to flow out of the regions of higher liquid content to the dryer regions due to surface tension force.

The second stage features the moisture sorption of fibers, which is relatively slow and takes a few minutes to a few hours to complete. In this period, water sorption into the fibers takes place as the water vapor diffuses into the fabric, which increases the relative humidity at the surfaces of fibers. After liquid water diffuses into the fabric, the surfaces of the fibers are saturated due to the film of water on them, which again will enhance the sorption process. During these two transient stages, heat transfer is coupled with the four different forms of liquid transfer due to the heat released or absorbed during sorption/desorption and evaporation/condensation. Sorption/desorption and evaporation/condensation, in turn, are affected by the efficiency of the heat transfer. For instance, sorption and evaporation in thick cotton fabric take a longer time to reach steady states than in thin cotton fabrics.

Finally, the third stage is reached as a steady state, in which all four forms of moisture transport and the heat transfer process become steady, and the coupling effects among them become less significant. The distributions of temperature, water vapor concentration, fiber water content, liquid volume fraction, and evaporation rate become invariant in time. With the evaporation of liquid water at the upper surface of the fabrics, liquid water is drawn from capillaries to the upper surface [12].

8.3   THERMAL CONDUCTIVITY

One way of expressing the insulating performance of a textile is to quote "effective thermal conductivity". Here the term "effective" refers to the fact that conductivity is calculated from the rate of heat flow per unit area of the fabric divided by the temperature gradient between opposite faces. It is not true condition, because heat transfer takes place by a combination of conduction through fibers, air, and infrared radiation.

If moisture is present, other mechanisms may be also involved. Research on the thermal resistance of apparel textiles has established that the thermal resistance of a dry fabric or one containing very small amounts of water depends on its thickness and to lesser extent on fabric construction and fiber conductivity. Indeed, measurements of effective thermal conductivity by standard steady-state methods show that differences between fabrics and mainly attributable to thickness. Despite these findings, consumers continue to regard wool as "warmer" than other fibers, and show preference for wearing wool garments in cold weather, particularly when light rain or sea spray is involved.

Meanwhile, the effective thermal conductivities of fabrics can be studied for varying regains. Regain is the mass of water present expressed as a percentage of the dry weight of the material. The effective thermal conductivities for porous acrylic, polypropylene, wool, and cotton is shown in Figure 1.

The curves indicate that changes in "effective thermal conductivity" with increasing regain are not linear but can be explained in terms of water within the fibers of fabrics with regain [13].

Figure 2 presents the various phases diagrammatically. When fabrics containing water are subjected to a temperature gradient, three different modes of heat flow can be distinguished:

- The presence of condensed water.
- Vapor transport.
- Condensation.

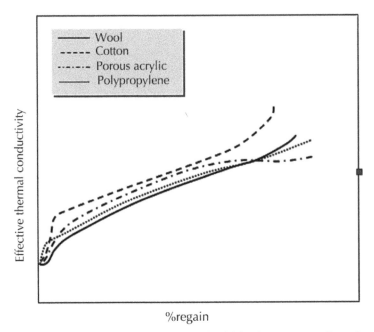

FIGURE 1   Comparison of effective thermal conductivities for porous acrylic, polypropylene, wool, and cotton.

Fiber sorption properties influence the heat and mass transfer up to the point when the rate of increased conductivity with regain is low in the curves, and then all fiber types behave similarly. Generally, heat transfer increases with increasing regain, but in this initial regain the rise is most pronounces for the nonabsorbent polypropylene. The fiber with the lowest effective conductivity over the regain 0–200% regain is wool, an effect that is especially pronounced in the region of low regains from zero to saturation. This is mostly influenced by fiber sorption properties. Low regains are most common in real wear situation. This is mostly influenced by fiber sorption properties. Low regains are most common in real wear situations.

This explains the popular association between wool and warmth in situations such as yachting, where the garment will very likely become wet. Cotton fabric has the highest effective thermal conductivity for almost the whole regain [14].

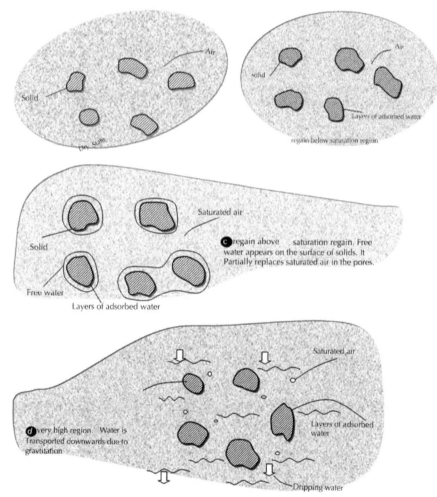

FIGURE 2   (a) Cross sections of nonabsorbent porous textile at different regains.

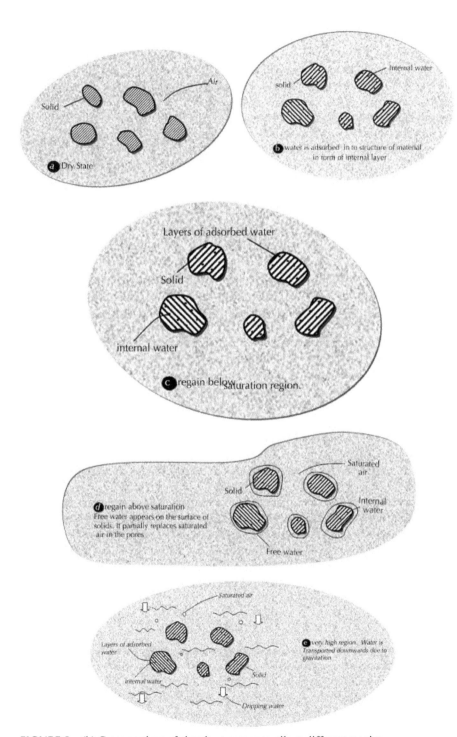

FIGURE 2   (b) Cross sections of absorbent porous textile at different regains.

8.4   TRANSPORT PHENOMENA

Fabrics to protect human body are, in most cases used under nonequilibrium conditions: therefore, characteristics of fabrics under nonisothermal and nonequilibrium conditions are important in evaluating overall performance. Furthermore, in colder environments, layered fabrics rather than a single fabric are used in most cases. Under such conditions, the two most important characteristics of fabrics are water vapor and heat transport. However, the water vapor transport may not influence significantly by surface characteristics the hydrophilic or hydrophobic nature of fabrics. On the other hand, when liquid water contacted a fabric, such as in the case of sweating, the surface wetability of fabric play a dominant role in determining the water vapor transport through layered fabrics.

In such a case, the wicking characteristics, which determines how quickly and how widely liquid water spreads out laterally on the surface of or within the matrix of the fabric, determines the overall water vapor transport rate through the layered fabrics [15].

It should be noted that the overall water vapor and heat transport characteristics of a fabric should depend on other factors such as the water vapor absorbability of the fibers, the porosity, density, and thickness of the fabric, and so on.

Moreover, transport phenomena for the sweat case are much more complicated than the water vapor case because wetting of the surface by liquid water precedes water wetting of the surface by liquid water precedes water vapor transmission. Note that there is an important difference in water absorbing characterizing of wool and cotton, although both fibers have relatively high water vapor absorption rates. Because of the hydrophobic surface of wool fibers, a liquid droplet in contact with a wool fabric does not spread out laterally within a fabric layer. The water vapor transport rate, in the sweat case, can be indicated by the size of liquid water spread out on the surface or within a fabric matrix.

Moreover, the term "breathable" implies that the fabric is actively ventilated. This is not the case. Breathable fabrics passively allow water vapor to diffuse through them yet still prevent the penetration of liquid water. Production of water vapor by the skin is essential for maintenance of body temperature. The normal body core temperature is 37°C, and skin temperature is between 33 and 35°C, depending on conditions. If the core temperature goes beyond critical limits of about 24 and 45°C then death results. The narrower limits of 34 and 42°C can cause adverse effects such as disorientation and convulsions. If the sufferer is engaged in a hazardous pastime or occupation then this could have disastrous consequences [16].

8.5   FACTORS INFLUENCING THE COMFORT

As it was mentioned earlier, if liquid water (sweat) cannot be dissipated quickly, the humidity of the air in the space between the skin and the fabric that contacts with the skin rises. This increased humidity prevents rapid evaporation of liquid water on the skin and gives the body the sensation of "heat" that triggered the sweating in the first place. Consequently, the body responds with increased sweating to dissipate excess thermal energy. Thus a fabric's inability to remove liquid water seems to be the major factor causing uncomfortable feeling for the wearer.

Scientists conducted wearer trails for shirts made of various fibers. They concluded that the largest factor that influences wearing comfort was the ability of fibers to absorb water, regardless of whether fibers were synthetic or natural. All of these studies indicate that the transient state phenomenon responding to the physiological demand to cause sweating is most relevant to comfort or discomfort associated with fabrics. It is important to point out that a highly water absorbing fabric placed in the first layer keeps the partial pressure of water-vapor near the skin low, which helps to dissipate water at the skin surface, although the vapor transport rate is smaller than for non-absorbing fabrics. In other words, the dissipation of water by means of absorption by fabrics appears to be more efficient way to keep water-vapor pressure near the skin low than dissipation by permeation through fabrics. Highly water absorbing fabrics raise the temperature of the air space near the skin. The temperature rise will further decrease relative humidity: however, the higher temperature may or may not be desirable depending on environmental conditions [17].

8.6   INTERACTION OF MOISTURE WITH POROUS TEXTILE

Trying to stay warm and dry while active outdoors in winter has always been a challenge. In the worst case, an individual exercises strenuously, sweats profusely, and then rests. During exercise, liquid water accumulates on the skin and starts to wet the clothing layers above skin. Some of the sweat evaporates from both the skin and the clothing. Depending on the temperature and humidity gradient across the clothing, the water vapor either leaves the clothing or condenses and freezes somewhere in its outer layers.

When one stop exercising and begins to rest, active sweating soon ceases, allowing the skin and clothing layers eventually dry. During this time, however, the heat loss from body can be considerable. Heat is taken from the body to evaporate the sweat, both that on the skin and that in the clothing. The heat flow from the skin through the clothing can be considerably greater when the clothing is very wet, since water decreases clothing's thermal insulation. This post-exercise chill can be exceedingly comfortable and can lead to dangerous hypothermia.

A dry layer next to the skin is more comfortable than a wet one. If one can wear clothing next to the skin that does not pick up any moisture, but rather passes it through to a layer away from the skin, heat loss at rest will be reduced. For such reasons, synthetic fibers have gained popularity with winter enthusiasts such as hikers and skiers [18].

Advertising the popular press would have us believe that synthetic materials pick up very little moisture, dry quickly, and so leave the wearer warm and dry. In contrast, warnings are given against wearing cotton or wool next to skin, since these fibers absorb sweat and so "lower body temperature". A further property credited to synthetics, in particular polypropylene, is that they wick water away from the skin, leaving one dry and comfortable.

In the early fifties, when synthetic fibers such as nylon and the acrylics were first coming onto the consumer market, Fourt et al. (1951) and Coplan (1953) compared the water absorption and drying properties of these "miracle" fibers with those of conventional wool and cotton. Forty-five years later, the water absorption and drying properties of synthetics were compared with natural fibers and it was found that all fabrics pick up water, and the time they take to dry is proportional to the amount of water they initially pick up.

It was also found that properties relevant to clothing on an exercising person, that is, the energy required to evaporated water from under and through a dry fabric or to dry a wet fabric and layer-to-layer wicking.

Researchers compared the heat exchange and thermal insulation of two ensembles, one made from wool, the other from nylon, worn by subjects who exercised either lightly (dry condition) or strenuously (wet condition) for 60 min, then rested 60 min [19].

They found that there was a significant difference in the physiological and subjective responses between dry and wet conditions, but not between the two fiber types. Further, there was no significant difference between the ratings of temperature and humidity sensations for the wool and nylon garments. The wool garment picked up more water than the nylon garment (245 g versus 198 g) for the wet condition. However, the wool fabric may have been slightly thicker than the nylon fabric, since it was reported to have a slightly greater thermal resistance and would therefore hold more water.

8.7   MOISTURE TRANSFER IN POROUS TEXTILES

In nude man any increase of sweating is immediately accompanied by an increase in heat loss due to evaporation. Similarly, any decrease in sweating is immediately accompanied by a decrease in heat loss. Thus, nude man has a control of his heat loss which has no appreciable time lag. This is shown diagrammatically in Figure 3.

In this figure, time is plotted as abscissa and rate of heat produced or lost as ordinate. To maintain perfect heat balance and a constant temperature, heat loss should equal heat production so that the heat production and heat loss curves should be the same. Suppose a man is initially at rest with a low heat production and a like heat loss as represented by the solid line in period A. When he exercises and produce more heat, the heat loss should rise as represented by the solid line in period B. Again, when he returns to the resting condition, period C, heat loss should return to the solid base line [20].

If sweating is the mechanism bringing about increasing heat loss but evaporation is limited, the increased heat loss might only be sufficient to match an increased heat production represented by the dashed line in period B. The position of this dashed line will depend on the equilibrium vapor transfer characteristics of the clothing. If, however, the hypothetical man is clothed in absorbent clothing, some of the sweat initially evaporated on the skin at the start of the exercise period will be absorbed by the clothing and its heat of absorption will appear in the clothing as sensible heat. This source of sensible heat will temporarily reduce the heat loss so that it follows the dotted line. Eventually, new equilibrium moisture content will be established and the dotted and dashed lines will coincide. When exercise and sweating stop, period C, moisture accumulated in the clothing will be desorbed or evaporated and tend to cool the clothing and the man wearing it. Thus, there is a time lag, and the heat loss curve will tend to follow the dotted curve during the after exercise period. Since in Figure 3 heat loss per unit time is plotted against time, the area between the dotted line and the solid line represents an amount of heat, as distinguished from rate of heat loss, which can be regarded as a quantitative value of after exercise chill.

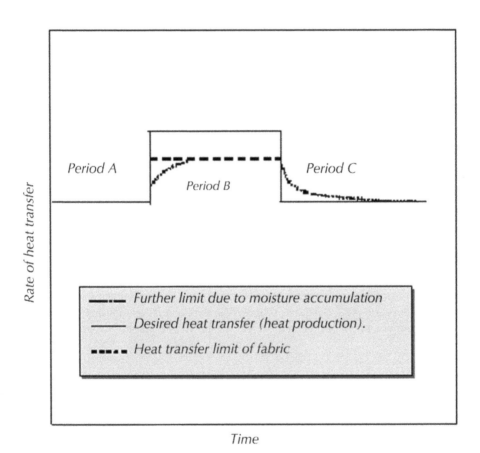

FIGURE 3   Rate of heat transfer versus time.

It should be noted that the moisture contained in the clothing need not be only that which is collected by absorption. It is also possible in cold damp or extreme cold environments that sweat which is evaporated at the skin will recondense when it reaches colder layers of clothing. Alternatively, the sweat rate may be so high that some of it will not evaporate from the skin. In nude man this drips off, but in clothed man it is blotted up by clothing to evaporate after sweating ceases [21].

Meanwhile, measurements of water vapor permeability of woven fabrics have indicated that in the lower ranges of fabric density, the main path of water vapor transfer is through the air spaces between fibers and yarns. This covers the densities characteristic of most apparel fabrics made from staple fibers, although filament yarn fabrics may be woven to higher densities in which the kind of fiber itself in the passage of water vapor, it is necessary to account for the water vapor passage through air spaces.

8.8 WATER VAPOR SORPTION MECHANISM

Scientists proposed a mathematical model for describing heat and moisture transfer in fabric, as shown in Equations 1 and 2, and the further analyzed the model in 1948:

$$\varepsilon \frac{\partial C_a}{\partial t} + (1 - \varepsilon) \frac{\partial C_f}{\partial t} = \frac{D_a \varepsilon}{\tau} \frac{\partial^2 C_a}{\partial x^2} \tag{1}$$

$$C_v \frac{\partial T}{\partial t} - \lambda \frac{\partial C_f}{\partial t} = K \frac{\partial^2 T}{\partial x^2} \tag{2}$$

In these equations, both C_v and λ are functions of the concentration of water absorbed by the fibers. Most textile fibers have very small diameters and very large surface/volume ratios. The assumption in the second equation of instantaneous thermal equilibrium between the fibers and the gas in the interfiber space does not therefore lead to appreciable error. The two equations in the model are not linear and contain three unknown, that is C_f, T, and C_a. A third equation should be established appropriately in order to solve the model. Scientists derived a third equation to obtain an analytical solution by assuming that C_f is linearly dependent on T and C_a, and that fibers reach equilibrium with adjacent air instantaneously. Considering the two-stage sorption process of wool, Scientists proposed an exponential relationship to describe the rate of water content change in the fibers, as shown in Equations (3) and (4):

$$\frac{1}{\varepsilon} \frac{\partial C_f}{\partial t} (H_a - H_f) \gamma \tag{3}$$

where

$$\gamma = k_1 (1 - \exp(k_2 |H_a - H_f|)) \tag{4}$$

and k_1 and k_2 are adjustable parameters that are evaluated by comparing the prediction of the model and measured moisture content of the fabric.

Researchers reported a numerical model describing combined heat and water vapor transport through clothing. The assumptions in his model do not allow for the complexity of the moisture sorption isotherm and the sorption kinetics of fibers. Scientists presented two mathematical models to simulate the interaction between moisture sorption by fiber and moisture flux through the air spaces of a fabric. In the first model, they considered diffusion within the fiber to be so rapid that the fiber moisture content is always in equilibrium with the adjacent air. In the second model, they assumed that the sorption kinetics of the fiber follows Fickian diffusion. Their model neglected the effect of heat of sorption behavior of the fiber. Scientists developed a new sorption equation that takes into account the two-stage sorption kinetics of wool fibers, and incorporated this with more realistic boundary conditions to simulate the sorption behavior of wool fabrics. They assumed that water vapor uptake rate of fiber consists of a two components associated with the two stages of sorption [22].

The first stage is represented by Fickian diffusion with a constant coefficient. Second-stage sorption is much slower than the first and follows an exponential relationship. The relative contributions of the two stages to the total uptake vary with the sorption stage and the initial regain of the fibers. Thus, the sorption rate equation can be written as:

$$\frac{\partial C_f}{\partial t}(1-p)R_1 + pR_2 \tag{5}$$

where R_1 is the first-stage sorption rate, R_2 is the second-stage rate sorption rate, and p is a proportional of uptake in the second stage. Equation (5) assumes that the sorption rate is a linear average of the first and second sorption rates. The first-stage sorption rate R_1 can be derived using Crank's (1995) truncated solution which may lead to a corresponding algorithm that needs a strict time striction and hence long computation times.

The second-stage sorption rate R_2, which relates local temperature, humidity, and the sorption history of the fabric, is assumed to have the following form:

$$R_2(x,t) = s_1 sign(H_a(x,t) - H_a(x,t) - H_f(x,t)$$
$$\times \exp\left(\frac{s_2}{|H_a(x,t) - H_f(x,t)|}\right) \tag{6}$$

where s_1 and s_2 are constants. No values for s_1 and s_2 have been reported in the literature for any textile fibers. This is also an empirical equation that has an unclear physical meaning, which makes it inconvenient to predict and simulate heat and moisture transport in a fabric. These equations were improved substantially by researchers. The numerical values and approximate relationships they used are listed in Table 3. They assumed that moisture sorption by a wool fiber can be generally described by a uniform diffusion equation for both stages of sorption:

$$\frac{\partial C_f(x,r,t)}{\partial t} = \frac{1}{r}\frac{\partial}{\partial r}\left(rD_f(x,t)\frac{\partial C_f(x,r,t)}{\partial r}\right) \tag{7}$$
$$C_{fs}(x,R_f,t) = f(H_a(x,t),T(x,t))$$

where $D_f(x,t)$ are the diffusion coefficients that have different presentations at different stages of sorption, and x is the coordinate of a fiber in the given fabric. The boundary condition is determined by the relative humidity of the air surrounding a fiber at x. In a wool fabric, $D_f(x,t)$ is a function of $W_c(x,t)$, which depends on the sorption time and the fiber location.

TABLE 3 Numerical values of wool and physical properties.

Parameters	Initial values	Mathematical relationship
Thermal conductivity of fabric (KJ/mK)	$3.8493e^{-2}$	$(38.493 - 0.72W_c + 0.113W_c^2$ $-0.002W_c^3)10^{-3}$
Volumetric heat capacity of fabric	1609.7	373.3+4661 +4.221 T
Diffusion coefficient of fiber	2.4435	1.0637 arc tan(1541.1933) (3600/)
Diffusion coefficient of water vapor in fabric (	1.91	_____
Heat of sorption or desorption of water by fibers	4124.5	1602.5exp(−11.72) +2522
Porosity of fabric	0.925	_____
Density of fabric	1330	_____
Radius of wool fiber (m)		_____
Mass transfer coefficient (m/s)	0.137	_____
Heat transfer coefficient	99.4	_____

=Water content of the fibers in the fabric

8.9 MODELING

The fabric model simulates the transport of a liquid and vapor-phase fluid that can undergo phase change (e.g., water) and an inert gas (air) in a textile layer. Several new models and capabilities were added to a standard commercial CFD code.

These capabilities include:
- Vapor phase transport (variable permeability).
- Liquid phase transport (wicking).
- Fabric property dependence on moisture content.
- Vapor/liquid phase change (evaporation/condensation).
- Sorption to fabric fibers.

In the fabric, transport equations are derived for mass, momentum, and energy in the gas and liquid phases by volume-averaging techniques. Definitions for intrinsic phase average, global phase average, and spatial average for porous media are those given by Whitaker (1998). Since the fabric porosity is not constant due to changing amounts of liquid and bound water, the source term for each transport equation includes quantities that arise due to the variable porosity. These equations are summarized in general form:

- Gas phase continuity Equation:

$$\frac{\partial}{\partial t}\left((1-\varepsilon_{ds})\rho_{\gamma}\right)+\nabla.\left(\rho_{\gamma}v_{\gamma}\right)=S_{\gamma} \tag{8}$$

$$S_{y}=m'_{sv}+m'_{lv}+\frac{\partial}{\partial t}\left(\left(\varepsilon_{bl}+\varepsilon_{\beta}\right)\rho_{\gamma}\right) \tag{9}$$

- Vapor continuity Equation:

$$\frac{\partial}{\partial t}\left((1-\varepsilon_{ds})\rho_{\gamma}m_{\nu}\right)+\nabla.\left(\rho_{\gamma}m_{\nu}v_{\gamma}\right)=\nabla.\left\{\rho_{\gamma}D_{eff}\nabla(m_{\nu})\right\}+S_{\nu} \tag{10}$$

$$S_{\nu}=m'''_{sv}+m'''_{lv}+\frac{\partial}{\partial t}\left\{\left(\varepsilon_{bl}+\varepsilon_{\beta}\right)\rho_{\gamma}m_{\nu}\right\} \tag{11}$$

$$m_{\nu}=\frac{\rho_{\nu}}{\rho_{\gamma}} \tag{12}$$

- Gas phase momentum equation:

$$\frac{\partial}{\partial t}\left(\rho_{\gamma}v_{\gamma}\right)+\nabla.\left(\rho_{\gamma}v_{\gamma}\right)=\nabla.\left\{\mu_{\gamma}\nabla(v_{\gamma})\right\}-\nabla p_{\gamma}+S_{\gamma} \tag{13}$$

$$S_{\gamma}=-v_{\gamma}\frac{\mu_{\gamma}}{K_{\gamma}k_{\gamma}} \tag{14}$$

- Liquid transport:

$$\frac{\partial}{\partial t}\left[(1-\varepsilon_{ds})\rho_{\gamma}s\right]=\nabla.\left(-\frac{k_{\beta}K_{\beta}}{\mu_{\beta}}\frac{\partial P_{c}}{\partial s}\right)\nabla s+S_{l} \tag{15}$$

$$S_{l}=-\nabla.\left(\frac{k_{\beta}K_{\beta}}{\mu_{\beta}}\rho_{\beta}g\right)-\frac{m'''_{ls}}{\rho_{\beta}}-\frac{m'''_{lv}}{\rho_{\beta}}+\frac{\partial}{\partial t}\left[(\varepsilon_{bl})s\right]+\frac{\partial}{\partial t}\left[(1-\varepsilon_{ds})(\rho_{\gamma}-1)s\right] \tag{16}$$

$$\frac{\partial}{\partial t}\left[(1-\varepsilon_{ds})\rho_{\gamma}h_{\gamma}+\varepsilon_{ds}\rho_{ds}h_{ds}\right]+\nabla.\left(\rho_{\gamma}v_{\gamma}h_{\gamma}\right)+\nabla.\left(\rho_{\beta}v_{\beta}\left(h_{\nu}-\Delta h_{\nu}\right)\right)$$
$$=\nabla.\left[k_{eff}\nabla(T)\right]+\nabla.\left(-h_{a}J_{a}-h_{\nu}J_{\nu}\right)+S_{T} \tag{17}$$

$$S_T = \frac{\partial}{\partial t}\left[\left(\varepsilon_\beta + \varepsilon_{bl}\right)\rho_\gamma h_\gamma - \varepsilon_\beta \rho_\beta \left(h_\nu - \Delta h_\nu\right) - \varepsilon_{bl}\rho_\beta \left(h_\nu - \Delta h_\nu - Q\right)\right] \tag{18}$$

$$\nu_\beta = \frac{k_\beta K_\beta}{\mu_\beta}\left[\frac{\partial P_c}{\partial s}\nabla s + \rho_\beta g\right] \tag{19}$$

8.10 CONCLUSION

In summary, due to the intensive body activity, the wearer perspires, and the cloth worn next to skin will get wet. These moisture fabrics reduce the body heat and make the wearer to become tired. So the cloth worn next to the skin should assist for the moisture release quickly to the atmosphere. The fabric worn next to the skin should have two important properties. The initial and for most property is to evaporate the perspiration from the skin surface and the second property is to transfer the moisture to the atmosphere and make the wearer to feel comfort. Diffusion and wicking are the two ways by which the moisture is transferred to the atmosphere. These two are mostly governed by the fiber type and fabric stricture. The air flow through the fabric makes the moisture to evaporate to the atmosphere. The capillary path plays a vital role in the transfer of moisture and this depends on the wicking behavior of the fabric. In the development of protective clothing and other textiles, modeling offers a powerful companion to experiments and testing.

It should be noted that condensation occurs when the vapor density of the steam is higher that its saturation vapor density. The condensation rate is proportional to the vapor density difference between that in the gas phase and that at the condensing surface. The relative hygrometry is quite different due to the action (water vapor pressure) of the airs on each side of the fabric (Figure 4). Sorption and desorption have not opposite kinetics, the former faster in temperature and in charge of humidity during

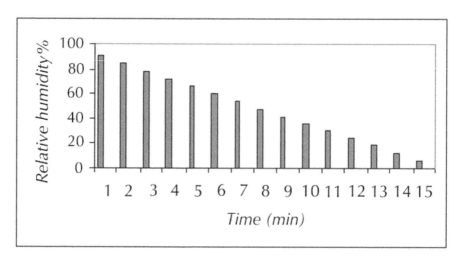

FIGURE 4   Relative hygrometry of wool fiber during desorption.

the first minutes, the later more completer in discharge of humidity after a long time (Figure 5). Figure 6 shows experimental results of hygrometry for the internal and external surfaces of wool fabric (between skin and fabric). Meanwhile, the internal gap has a considerable effect on the moisture transmission rate. The internal air gap has been identified as being a source of potential errors in most experimental works due to its changing resistance (Figure 7). Figure 8 shows the effect of thickness on the amount of water can be held in a fabric.

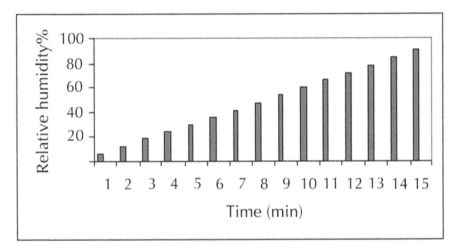

FIGURE 5   Relative hygrometry of wool fiber during sorption.

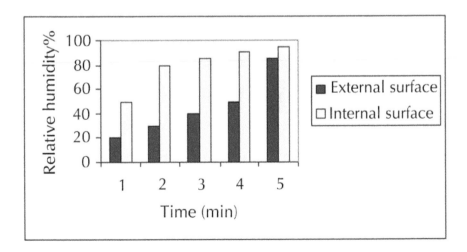

FIGURE 6   Experimental results of hygrometry for the internaland external surfaces of wool fabric (between skin and fabric).

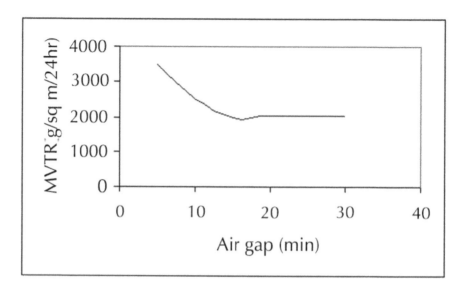

FIGURE 7 Effect of the internal air gap size on the moisture vapor transmission (MVTR).

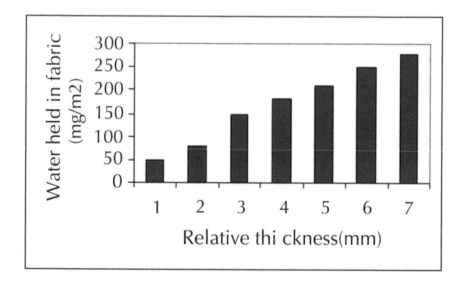

FIGURE 8. Thickness *versus* the amount of water held in fabric.

KEYWORDS

- **Fickian diffusion**
- **Moisture vapor transmission**
- **Thermal conductivity**
- **Thermal environment**
- **Water vapor resistance**

REFERENCES

1. Incropera, F. P. and Dewitt, D. P. *Fundamentals of Heat and Mass Transfer*. Second ed., Wiley, New York (1985).
2. Ghali, K., Jones, B., and Tracy, E. Modeling Heat and Mass Transfer in Fabrics. *Int. J. Heat Mass Transfer*, **38**(1), 13–21 (1995).
3 .Flory, P. J. *Statistical Mechanics of Chain Molecules*. Interscience Pub., New York (1969).
4. Hadley, G. R. Numerical Modeling of the Drying of Porous Materials. *In Proceedings of The Fourth International Drying Symposium, Vol. 1*, pp. 151–158 (1984).
5. Hong, K., Hollies, N. R. S., and Spivak, S. M. Dynamic Moisture Vapor Transfer through Textiles, Part I Clothing Hygrometry and the Influence of Fiber Type. *Textile Res. J.*, **58**(12), 697–706 (1988).
6. Chen, C. S. and Johnson, W. H. Kinetics of Moisture Movement in Hygroscopic Materials, In Theoretical Considerations of Drying Phenomenon. *Trans. ASAE.*, **12**, 109–113 (1969).
7. Barnes, J. and Holcombe, B. Moisture Sorption and Transport in Clothing during Wear. *Textile Res. J.*, **66**(12), 777–786 (1996).
8. Chen, P. and Pei, D. A. Mathematical Model of Drying Process. *Int. J. Heat Mass Transfer*, **31**(12), 2517–2562 (1988).
9. Davis, A. and James, D. Slow Flow through a Model Fibrous Porous Medium. *Int. J. Multiphase Flow*, **22**, 969–989 (1996).
10. Jirsak, O., Gok, T., Ozipek, B., and Pau, N. Comparing Dynamic and Static Methods for Measuring Thermal Conductive Properties of Textiles. *Textile Res. J.*, **68**(1), 47–56 (1998).
11. Kaviany, M. *Principle of Heat Transfer in Porous Media*. Springer, New York (1991).
12. Jackson, J. and James, D. The Permeability of Fibrous Porous Media. *Can. J. Chem. Eng.*, **64**, 364–374 (1986).
13. Dietl, C., George, O. P., and Bansal, N. K. Modeling of Diffusion in Capillary Porous Materials During the Drying Process. *Drying Technol.*, **13**(1, 2), 267–293 (1995).
14. Ea, J. Y. *Water Vapor Transfer in Breathable Fabrics for Clothing*. PhD thesis, University of Leeds (1988).
15. Haghi, A. K. Moisture permeation of clothing, *JTAC*, **76**, 1035–1055(2004).
16. Haghi, A. K. Thermal analysis of drying process, *JTAC*, **74**, 827–842(2003).
17. Haghi, A. K. *Some Aspects of Microwave Drying*. The Annals of Stefan cel Mare University, Year VII, No. 14, pp. 22–25 (2000).
18. Haghi, A. K. A *Thermal Imaging Technique for Measuring Transient Temperature Field an Experimental Approach*. The Annals of Stefan cel Mare University, Year VI, No. 12, pp 73–76 (2000).
19. Haghi, A. K. Experimental Investigations on Drying of Porous Media using Infrared Radiation. *Acta Polytechnica*, **41**(1), 55–57 (2001).
20. Haghi, A. K. A Mathematical Model of the Drying Process, *Acta Polytechnica*, **41**(3) 20–23 (2001).

21. Haghi, A. K. Simultaneous Moisture and Heat Transfer in Porous System. *Journal of Computational and Applied Mechanics*, **2**(2), 195–204 (2001).
22. .Haghi, A. K. A Detailed Study on Moisture Sorption of Hygroscopic Fiber. *Journal of Theoretical and Applied Mechanics*, **32**(2) 47–62 (2002).

9 Heat Flow in Non-homogeneous Material

CONTENTS

9.1 INTRODUCTION

Wood is a hygroscopic, porous, anisotropic, and non-homogeneous material. After log sawing, the lumber contains liquid water in fiber cavities (capillary water) and bound water inside the fiber wall (hygroscopic water). Porosity refers to volume fraction of void space. This void space can be actual space filled with air or space filled with both water and air. Capillary-porous materials are sometimes defined as those having pore diameter less than 10^{-7} m. Capillary porous materials were defined as those having a clearly recognizable pore space. In capillary porous material, transport of water is a more complex phenomenon. In addition to molecular diffusion, water transport can be due to vapor diffusion, surface diffusion, Knudsen diffusion, capillary flow, and purely hydrodynamic flow. In hygroscopic materials, there is large amount of physically bound water and the material often shrinks during heating.

9.2 COMPUTER MODELS FOR HEAT FLOW IN NON-HOMOGENEOUS MATERIAL

In hygroscopic materials there is a level of moisture saturation below which the internal vapor pressure is a function of saturation and temperature. These relationships are called equilibrium moisture isotherms. Above this moisture saturation, the vapor pressure is a function of temperature only (as expressed by the Clapeyron equation) and is independent of the moisture level. Thus, above certain moisture level, all materials behave non-hygroscopic [1].

Green wood contains a lot of water. In the outer parts of the stem, in the sapwood, spruce, and pine have average moisture content of about 130%, and in the inner parts, in the heartwood, the average moisture content is about 35%. Wood drying is the art of getting rid of that surplus water under controlled forms. It will dry to an equilibrium moisture content of 8–16% fluid content when left in air which improves its stability, reduces its weight for transport, prepares it for chemical treatment or painting and improves its mechanical strength.

Water in wood is found in the cell cavities and cell walls. All void spaces in wood can be filled with liquid water called free water. Free water is held by adhesion and surface tension forces. Water in the cell walls is called bound water. Bound water is held by forces at the molecular level. Water molecules attach themselves to sites on the cellulose chain molecules. It is an intimate part of the cell wall but does not alter the chemical properties of wood. Hydrogen bonding is the predominant fixing mechanism. If wood is allowed to dry, the first water to be removed is free water. No bound water is evaporated until all free water has been removed. During removal of water, molecular energy is expended. Energy requirement for vaporization of bound water is higher than free water. Moisture content at which only the cell walls are completely saturated (all bound water) but no free water exists in all lumens is called the fiber saturation point (FSP). Typically the FSP of wood is within the range of 20–40% moisture content depending on temperature and wood species. Water in wood normally moves from high to low zones of moisture content. The surface of the wood must be drier than the interior if moisture is to be removed. Drying can be divided into two phases: movement of water from the interior to the surface of the wood and removal of water from the surface. Water moves through the interior of the wood as a liquid or water vapor through various air passageways in the cellular structure of wood and through the cell walls [2].

Drying is a process of simultaneous heat and moisture transfer with a transient nature. The evolution process of the temperature and moisture with time must be predicted and actively controlled in order to ensure an effective and efficient drying operation. Lumber drying can be understood as the balance between heat transfer from air flow to wood surface and water transport from the wood surface to the air flow. Reduction in drying time and energy consumption offers the wood industries a great potential for economic benefit. In hygroscopic porous material like wood, mathematical models describing moisture and heat movements may be used to facilitate experimental testing and to explain the physical mechanisms underlying such mass transfer processes. The process of wood drying can be interpreted as simultaneous heat and moisture transfer with local thermodynamic equilibrium at each point within the timber. Drying of wood is in its nature an unsteady-state nonisothermal diffusion of heat and moisture, where temperature gradients may counteract with the moisture gradient [3].

9.3   SOME ASPECTS OF HEAT FLOW DURING DRYING PROCESS

9.3.1   Stages of Drying

First Stage

When both surface and core MC are greater than the FSP. Moisture movement is by capillary flow. Drying rate is evaporation controlled.

Second Stage

When surface MC is less than the FSP and core MC is greater than the FSP. Drying is by capillary flow in the core and by bound water diffusion near the surface as fiber saturation line recedes into wood, resistance to drying increases. Drying rate is controlled by bound water diffusion.

Third Stage

When both surface and core MC are less than the FSP. Drying is entirely by diffusion. As the MC gradient between surface and core becomes less, resistance to drying increases and drying rate decreases.

9.3.2   Capillary

Capillary pressure is a driving force in convective wood drying at mild conditions. The temperature is higher outside than inside. The moisture profile during convective drying is in the opposite direction, namely, the drier part is toward the exposed surface of wood. This opposite pattern of moisture and temperature profiles lead to the concept of the wet front that separates the outer area, where the water is bound to the cell wall, from the inner area, where free water exists in liquid and vapor form. A wet front that moves slowly from the surface toward the center of a board during convective drying leads to subsequent enhancement of the capillary transportation. Capillary transportation can then be justified due to the moisture gradients developed around that area. When the drying conditions are mild, the drying period is longer so the relative portion of the total moisture removal, due to the capillary phenomena, is high, and it seems that this is the most important mass transfer mechanism [4].

9.3.3   Bound Water Diffusion

Credible data on the bound water diffusion coefficient in wood and the boundary condition for the interface between moist air and wood surface are very important for accurate description of timber drying as well as for the proper design and use of products, structures and buildings made of wood already dried below the FSP. During the last century, two groups of methods for measuring the bound water diffusion coefficient in wood were developed. The first one, traditionally called the cup method, uses data from the steady-state experiments of bound water transfer and is based on Fick's first law of diffusion. Unfortunately, the method is not valid for the bound water diffusion coefficient determination in wood because it cannot satisfy the requirements of the boundary condition of the first kind and the constant value of the diffusion coefficient. The second group of methods is based on the unsteady-state experiments and Fick's second law of diffusion. The common name of this group is the sorption method and it was developed to overcome the disadvantages of the cup technique [5].

9.3.4　Diffusion

In solving the diffusion equation for moisture variations in wood, some authors have assumed that the diffusion coefficient depends strongly on moisture content, while others have taken the diffusion coefficient as constant. It has been reported that the diffusion coefficient is influenced by the drying temperature, density and moisture content of timber. The diffusion coefficient of water in cellophane and wood substance was shown to increase with temperature in proportion to the increase in vapor pressure of water. It is also observed that the diffusion coefficient decreased with increasing wood density. Other factors affecting the diffusion coefficient that are yet to be quantified are the species (specific gravity) and the growth ring orientation. Literature has suggested that the ratios of radial and tangential diffusion coefficients vary for different tree species. The radial diffusion coefficient of New Zealand Pinus radiate has been estimated to be approximately 1.4 times the tangential diffusion coefficient. It is observed that for northern red oak, the diffusion coefficient is a function of moisture content only. It increases dramatically at low moisture content and tends to level off as the FSP is approached.

In a 1D formulation with moisture moving in the direction normal to a specimen of a slice of wood of thickness 2a, the diffusion equation can be written as:

$$\frac{\partial(MC)}{\partial t} = \frac{\partial}{\partial X}\left(D\frac{\partial(MC)}{\partial X}\right)(0 < X < a, t > 0) \tag{1}$$

where, MC is moisture content, t is time, D is diffusion coefficient, and X is space coordinate measured from the center of the specimen.

The moisture content influences on the coefficient D only if the moisture content is below the FSP (typically 20–30% for softwoods):

$$D(u) = \begin{cases} f_D(u) & , u < u_{fsp} \\ \\ f_D(u_{fsp}) & , u \geq u_{fsp} \end{cases} \tag{2}$$

where, u_{fsp} denotes the FSP. and $f_D(u)$ is a function which expresses diffusion coefficient in moisture content, temperature and may be some other parameters of ambient air climate. The expression of $f_D(u)$ depends on variety of wood.

It was assumed that the diffusion coefficient bellow FSP can be represented by:

$$f_D(u) = A.e^{-\frac{5280}{T}} .e^{\frac{B.u}{100}} \tag{3}$$

where, T is the temperature in Kelvin, u is percent moisture content, A and B can be experimentally determined.

The regression equation of diffusion coefficient of Pinus radiata timber using the dry bulb temperature and the density is:

$$D(10^{-9}) = 1.89 + 0127 \times T_{DB} - 0.00213 \times \rho_s \quad (R^2{=}0.499) \tag{4}$$

The regression equations of diffusion coefficients below of Masson's pine during high temperature drying are:

$$D = 0.0046MC^2 + 0.1753MC + 4.2850\left(R^2 = 0.9391\right) \tag{5}$$

Tangential diffusion

$$D = 0.0092MC^2 + 0.3065MC + 4.9243\left(R^2 = 0.9284\right) \tag{6}$$

Radial diffusion

The transverse diffusion coefficient D can be expressed by the porosity of wood V, the transverse bound water diffusion coefficient D_{bt} of wood and the vapor diffusion coefficient D_v in the lumens:

$$D = \frac{\sqrt{\nu}D_{bt}D_v}{(1-\nu)\left(\sqrt{\nu}D_{bt} + \left(1-\sqrt{\nu}\right)D_v\right)} \tag{7}$$

The vapor diffusion coefficient D_v in the lumens can be expressed as:

$$D_v = \frac{M_w D_a P_s}{SG_d \rho_w RT_k} \cdot \frac{d\phi}{du} \tag{8}$$

Where M_w (kg/kmol) is the molecular weight of water.

$$D_a = \frac{9.2.10^{-9} T_k^{2.5}}{\left(T_k + 245.18\right)} \tag{9}$$

is the inter diffusion coefficient of vapor in air, is the nominal specific gravity of wood substance at the given bound water content.

$$SG_d = \frac{1.54}{(1+1.54u)} \tag{10}$$

$\rho_w = 103 \ kg/m^3$ is the density of water, $R = 8314.3$ kmol, K is the gas constant, T_k is the Kelvin temperature, Ψ is the relative humidity (%/100), and P_{sat} is saturated vapor pressure given by:

$$P_{sat} = 3390\exp\left(-1.74+0.0759T_C -0.000424T_C^2 +2.44.10^{-6}T_C^3\right) \tag{11}$$

The derivative of an relative humidity Ψ with respect to moisture content MC is given as:

$$MC = \frac{18}{w}\left(\frac{k_1 k_2 \psi}{1+k_1 k_2 \psi}+\frac{k_2 \psi}{1-k_2 \psi}\right) \tag{12}$$

where,

$$k_1 = 4.737+0.04773T_C -0.00050012T_C^2 \tag{13}$$

$$k_2 = 0.7059+0.001695T_C +-0.000005638T_C^2 \tag{14}$$

$$W = 223.4+.6942T_C +0.01853T_C^2 \tag{15}$$

The diffusion coefficient D_{bt} of bound water in cell walls is defined according to the Arrhenius equation as:

$$D_{bt} = 7.10^{-6} \exp\left(-E_b / RT_k\right) \tag{16}$$

where,

$$E_b = \left(40.195 - 71.179Mc + 291Mc^2 - 669.92Mc^3\right).10^6 \tag{17}$$

is the activation energy.
The porosity of wood is expressed as:

$$\nu = 1 - SG(0.667 + Mc) \tag{18}$$

Where specific gravity of wood SG at the given moisture content u is defined as:

$$SG = \frac{\rho_S}{\rho_w (1+ Mc)} = \frac{\rho_0}{\rho_w +0.883\rho_0 Mc} \tag{19}$$

Where, ρ_s is density of wood, ρ_0 is density of oven dry wood (density of wood that has been dried in a ventilated oven at approximately 104°C until there is no additional loss in weight).

9.3.5 Thermal Conductivity

Wood thermal conductivity (K_{wood}) is the ratio of the heat flux to the temperature gradient through a wood sample. Wood has a relatively low thermal conductivity due to its porous structure, and cell wall properties. The density, moisture content, and temperature dependence of thermal conductivity of wood and wood-based composites were demonstrated by several researchers. The transverse thermal conductivity can be expressed as:

$$K_{wood} = \left[SG \times (4.8 + 0.09 \times MC) + 0.57\right] \times 10^{-4} \frac{cal}{cm * Cs} \tag{20}$$

When moisture content of wood is below 40%.

$$K_{wood} = \left[SG \times (4.8 + 0.125 \times MC) + 0.57\right] \times 10^{-4} \frac{cal}{cm * Cs} \tag{21}$$

When moisture content of wood is above 40%.

The specific gravity and moisture content dependence of the solid wood thermal conductivity in the transverse (radial and tangential) direction is given by:

$$K_T = SG\left(K_{cw} + K_w.Mc\right) + K_a\nu \tag{22}$$

where,
SG = specific gravity of wood,
K_{cw} = Conductivity of cell wall substance (0.217 J/m/s/K),
K_w = conductivity of water (0.4 J/m/s/K),
K_a = conductivity of air (0.024 J/m/s/K),
Mc = moisture content of wood (fraction),
ν = porosity of wood.

The thermal conductivity of wood is affected by a number of basic factors: density, moisture content, extractive content, grain direction, structural irregularities such as checks and knots, fibril angle, and temperature. Thermal conductivity increases as density, moisture content, temperature, or extractive content of the wood increases. Thermal conductivity is nearly the same in the radial and tangential directions with respect to the growth rings.

The longitudinal thermal conductivity of solid wood is approximately 2.5 times higher than the transverse conductivity:

$$K_L = 2.5K_T \tag{23}$$

For moisture content levels below 25%, approximate thermal conductivity K across the grain can be calculated with a linear equation of the form:

$$K_{wood} = G(B+CM) + A \tag{24}$$

Where, SG is specific gravity based on oven dry weight and volume at a given moisture content MC (%) and A, B, and C are constants.

For specific gravity > 0.3, temperatures around 24°C, and moisture content values < 25%, $A = 0.01864$, $B = 0.1941$, and $C = 0.004064$ (with k in W/(m·K)). Equation (24) was derived from measurements made by several researchers on a variety of species [6].

9.4 DRYING RATES

During the early stages of drying the material consists of so much water that liquid surfaces exist and drying proceeds at a constant rate. Constant drying rates are achieved when surface free water is maintained and the only resistance to mass transfer is external to the wood. The liquid water moves by capillary forces to the surface in same proportion of moisture evaporation. Moisture movement across the lumber will depend on the wood permeability and the drying rate itself is controlled by external conditions in this period. Part of energy received by the surface increase temperature in this region and the heat transfer to the inner part of lumber starts. Since the moisture source for the surface is internal moisture, constant drying rates can only be maintained if there is sufficient moisture transport to keep the surface moisture content above the FSP. If this level is not maintained then some of the resistance to mass transfer becomes internal and neither the drying rate nor the surface temperature remains constant and drying proceeds to the falling rate period. As the lumber dries, the liquid water or wet line recedes into wood and the internal moisture movement involves the liquid flow and diffusion of water vapor and hygroscopic water. The effect of internal resistance on the drying rate increases. In the last phase (second falling rate period) there is no more liquid water in the lumber, and the drying rate is controlled only by internal resistance (material characteristics) until an equilibrium moisture content is reached [7-10].

A typical drying curve showing three stages of drying characteristic is illustrated in Figure 1.

Pang et al. proposed that the three drying periods (constant rate, first falling rate, and second falling rate) based on simulated drying of veneer be expressed by the following equations:

$$-\frac{d(MC)}{dt} = j_0 \quad \text{For } MC > {}^{M_{Cr1}} \tag{25}$$

$$-\frac{d(MC)}{dt} = A + B*MC \quad \text{For } M_{cr1} > MC > M_{cr2} \tag{26}$$

$$-\frac{d(MC)}{dt} = \frac{A + B*M_{cr2}}{M_{cr2} - M_e} * (MC - M_e) \quad \text{For MC} < M_{cr2} \qquad (27)$$

where, $\dot{J}_0$ is constant drying rate, M_{Cr1} is the first critical moisture content, M_{cr2} is the second critical moisture content, constants A and B also vary with wood thickness, wood density, and drying conditions [11, 12].

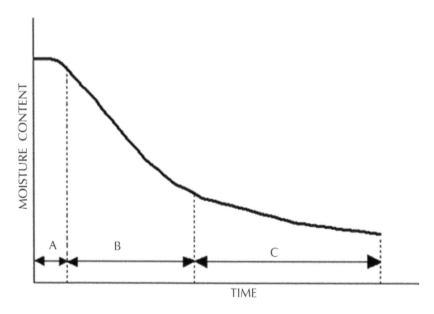

FIGURE 1 Drying characteristic of porous media: (a) constant rate region;(b) first falling rate region; (c) second falling rate region.

9.5 MOISTURE CONTENT AND PERMEABILITY

Moisture content of wood is defined as the weight of water in wood expressed as a fraction, usually a percentage, of the weight of oven dry wood. Moisture exists in wood as bound water within the cell wall, capillary water in liquid form and water vapor in gas form in the voids of wood. Capillary water bulk flow refers to the flow of liquid through the interconnected voids and over the surface of a solid due to molecular attraction between the liquid and the solid. Moisture content varies widely between species and within species of wood. It varies particularly between heartwood and sapwood. The amount of moisture in the cell wall may decrease as a result of extractive deposition when a tree undergoes change from sapwood to heartwood. The butt logs of trees may contain more water than the top logs. Variability of moisture content exists even within individual boards cut from the same tree. Green wood is often defined as freshly sawn wood in which the cell walls are completely saturated with water. Usually green wood contains additional water in the lumens. Moisture content at which both the cell lumens and cell walls are completely saturated with water is the maximum moisture content. An average green moisture content value taken from the Wood Handbook (Forest products society, 1999) of southern yellow pine (loblolly) is 33 and 110% for heartwood and sapwood, respectively. Sweetgum is 79 and 137% while yellow-poplar is 83 and 106% for heartwood and sapwood, respectively [13-15].

Permeability refers to the capability of a solid substance to allow the passage of gases or liquids under pressure. Permeability assumes the mass movement of molecules in which the pressure or driving force may be supplied by such sources as mechanically applied pressure, vacuum, thermal expansion, gravity, or surface tension. Under this condition, the permeability of wood is the dominant factor controlling moisture movement.

Fluid movement in wood is a very important process in wood products industries. An understanding of wood permeability is essential for determining lumber drying schedules for treating lumber and for producing high quality wood products. The flow of gas inside the wood particle is limited due to the fact that wood consists of a large number of clustered small pores. The pore walls act as barriers largely preventing convective flow between adjacent pores. The wood annular rings also act as barriers for flow in the radial direction which makes flow in the axial direction more favorable and giving a lower permeability in the radial direction than in the axial direction where the axial flow is regarded as flow parallel to the wood fiber grains and the radial flow as flow perpendicular to the wood grains. The permeability in the wood cylinder is therefore an important parameter for the velocity field in the wood. The dry wood radial permeability is 10,000 times lower than the dry wood axial permeability. The chemical composition of the wood/char structure also affects the permeability, where the permeability in char is in order of 1,000 times larger than for wood [16].

Longitudinal flow becomes important, particularly in specimens having a low ratio of length to diameter, because of the high ratio of longitudinal to transverse permeability. Longitudinal permeability was found to be dependent upon specimen length in the flow direction that is the decrease of specimen length appears result in greater permeability in less permeable species.

The effect of drying conditions on gas permeability and preservative treatability was assessed on western hemlock lumber. Although there were no differences in gas permeability between lumber dried at conventional and high temperatures, there were differences in preservative penetration. High temperature drying significantly reduced drying time, but did not appear to affect permeability or shell-to-core MC differences compared with drying at conventional temperature. Pits have a major influence on softwood permeability. Across pits can be impeded by aspiration or occlusion by deposition extractives on the membrane. Drying conditions can significantly affect pit condition, sometimes inducing aspiration that blocks both air and fluid flow. Pressure treatment is presumed to enhance preservative uptake and flow across pits, but the exact impact of pit condition (i.e., open or aspirated) is unknown. Drying conditions may also alter the state of materials deposited on pits, thereby altering the effects of pressure and perhaps the nature of preservative wood interactions. The latter effect may be especially important, since changes in wood chemistry could affect the rates of preservative fixation, which could produce more rapid preservative deposition on pit membranes that would slow further fluid ingress. The longitudinal permeability of the outer heartwood of each species also was determined to evaluate the effect of growth rate on the decrease in longitudinal permeability following sapwood conversion to heartwood. Faster diameter growth produced higher longitudinal permeability in the sapwood of yellow-poplar, but not in the sapwood of northern red oak or black walnut. Growth rate had no effect on either vessel lumen area percentage or decrease in longitudinal permeability in newly formed heartwood for all three species. Table 1 represents typical values for gas permeability. Values are given in orders of magnitude [17].

Darcy's law for liquid flow:

$$k = \frac{flux}{gradient} = \frac{V/(t \times A)}{\Delta P/L} = \frac{V \times L}{t \times A \times \Delta P} \tag{28}$$

where,

k = Permeability [cm^3] (liquid)/ (cm atm sec)].
V = Volume of liquid flowing through the specimen (cm^3).
t = Time of flow (sec).
A = Cross-sectional area of the specimen perpendicular to the direction of flow (cm^2).
ΔP = Pressure difference between ends of the specimen (atm).
L = Length of specimen parallel to the direction of flow (cm).

Darcy's law for gaseous flow:

$$K_g = \frac{V \times L \times P}{t \times A \times \Delta P \times \overline{P}} \tag{29}$$

where,

K_g = Superficial gas permeability [cm^3 (gas)/ (cm atm sec)].
V = Volume of gas flowing through the specimen (cm^3 (gas)).
P = Pressure at which V is measured (atm).
t = Time of flow (sec).
A = Cross-sectional area of the specimen perpendicular to the direction of flow (cm^2).
ΔP = Pressure difference between ends of the specimen (atm).
L = Length of specimen parallel to the direction of flow (cm).
$\overline{P}$ = Average pressure across the specimen (atm).

TABLE 1   Typical values for gas permeability.

Type of sample	Longitudinal gas permeability [cm^3 (gas)/(cm at sec)]
Red oak (R = 150 micrometers)	10,000
Basswood (R = 20 micrometers)	1,000

TABLE 1   *(Continued)*

Type of sample	Longitudinal gas permeability [*cm³* (gas)/(cm at sec)]
Maple, Pine sapwood, Coast Douglas-fir sapwood	100
Yellow-poplar sapwood, Spruce sapwood, Cedar sapwood	10
Coast Douglas-fir heartwood	1
White oak heartwood, Beech heartwood	0.1
Yellow-poplar heartwood, Cedar heartwood, Inland Douglasfir heartwood	0.01
Transverse Permeabilities (In approx. same species order as longitudinal)	0.001 – 0.0001

9.6   BASIC THEORETICAL CONCEPTS

9.6.1   Mass Conservation Equations

To simulate the heat and mass transport in drying, conservation equations for general non-hygroscopic porous media have been developed by Whitaker based on averaging procedures of all of the variables. These equations were further employed and modified for wood drying. Mass conservation equations for the three phases of moisture in local form are summarized in Equations (30–32).

Water vapor:

$$\frac{\partial}{\partial t}\left(\phi_g \rho_V\right) = -div\left(\rho_V V_V\right) + \dot{m}_{WV} + \dot{m}_{bV} \tag{30}$$

Bound water:

$$\frac{\partial}{\partial t}\left(\phi_s \rho_b\right) = -div\left(\rho_b V_b\right) + \dot{m}_{bV} + \dot{m}_{wb} \tag{31}$$

Free water:

$$\frac{\partial}{\partial t}\left(\phi_w \rho_w\right) = -div\left(\rho_w V_w\right) - \dot{m}_{wv} - \dot{m}_{wb} \tag{32}$$

where the velocity of the transported quantity is denoted by V_i, ρ_i is the density, and $\dot{m}_{ij}$ denotes the transition from phases i and j. From here on, the subscripts w, b, v, and s refer, respectively, to free water, bound water, water vapor, and the solid skeleton of wood. Denoting the total volume by V and the volume of the phase i by V_i, the volumetric fraction of this phase is:

$$\phi_i = \frac{V_i}{V} \tag{33}$$

with the geometrical constraint:

$$\phi_g + \phi_s + \phi_w = 1 \tag{34}$$

9.6.2   Generalized Darcy's Law

Darcy's law, by using relative permeabilities, provides expressions for the free liquid and gas phase velocities as follows:

$$\mathbf{v}^l = -\frac{K_l K_{rl}}{\mu_l} \nabla P_l \tag{35}$$

and

$$_v = -\frac{K_v K_{rv}}{\mu_v} \nabla P_v \tag{36}$$

where, K is the intrinsic permeability (m^2), K_r is the relative permeability, P is the pressure (Pa), and μ is the viscosity (Pa s).

9.6.3   External Heat and Mass Transfer Coefficients

The heat flux (q) and the moisture flux (N_v) are estimated by:

$$q = h\left(T_G - T_{surf}\right) \tag{37}$$

$$N_v = \psi K_0 \left(Y_{surf} - Y_G\right) = \beta\left(p_G^v - p_{ats}^v\right) \tag{38}$$

In which T_{surf}, Y_{surf} and p_s^v are respectively, the wood temperature, the air humidity, and the vapor partial pressure at the wood surface and, T_G, Y_G and p_G^v are the corresponding parameters in the air stream. The heat transfer coefficient is represented by h. The mass transfer coefficient is β when vapor partial pressure difference is taken as driving force and is k_0 when humidity difference is taken as the driving force with ψ being the humidity factor. The mass transfer coefficient related to humidity difference is a function of distance along the airflow direction from the inlet side. The heat transfer coefficient is correlated to the mass transfer coefficient, as shown by and can be calculated from it. The humidity coefficient φ has been found to vary from 0.70 to 0.76, depending on the drying schedules and board thickness [18, 19].

9.6.4   Moisture and Heat Balance Equations

For the moisture mass transfer and balance, the moisture loss from wood equals the moisture gain by the hot air, and the moisture transfer rate from the board is described by mass transfer coefficient multiplied by driving force (humidity difference, for example). These considerations yield:

$$-\frac{\partial}{\partial\tau}\left[MC.\rho_s.(1-\varepsilon)\right]=G.\frac{\partial Y}{\partial X}=\begin{cases}-\psi K_0.a.\left(Y_{surf}-Y_G\right)(condensation)\\ \psi K_0.a.f.\left(Y_{sat}-Y_G\right)(evaporation)\end{cases}$$ (39)

Where MC is the wood moisture content, ρ_s is the wood basic density, ε is the void fraction in the lumber stack, a is the exposed area per unit volume of the stack and G is the dry air mass flow rate. In order to solve the above equations, the relative drying rate (f) needs to be defined which is a function of moisture content.

For the heat transfer and balance, the energy loss from the hot air equals the heat gain by the moist wood. The convective heat transfer is described by product of heat transfer coefficient and the temperature difference between the hot air and the wood surface.

The resultant relationships are as follows:

$$\frac{\partial T_{wood}}{\partial\tau}=\frac{(1+\alpha_R-\alpha_{LS})}{\rho_s.(1-\varepsilon).C_{Pwood}}.\left[h.a.\left(T_G-T_{wood}\right)-G.\Delta H_{wv}.\frac{\partial Y_G}{\partial X}\right]$$

$$\frac{\partial T_G}{\partial X}=\frac{\left[h.a+G.C_{Pv}\frac{\partial Y_G}{\partial Z}\right].\left(T_G-T_{wood}\right)}{G.\left(C_{Pv}+Y_G.C_{Pv}\right)}$$ (40)

In the above equations, T_{wood} is the wood temperature, α_R and α_{LS} are coefficients to reflect effects of heat radiation and heat loss, C_{Pwood} is the specific heat of wood, and ΔH_{wv} is the of water evaporation. These equations have been solved to determine the changes of air temperature and wood temperature along the airflow direction and with time [20].

9.6.5 Energy Rate Balance on Drying Air and Wood
The energy rate balance (kW) of a drying air adjacent to the wood throughout the wood board can be represented as follows:

$$\frac{1}{2}V_a\rho_{a,mt}cp_{a,mt}\frac{dT_a}{dt}=\frac{1}{2}vA_{cs}cp_{a,mt}\left(T_{a,in}-T_{a,ex}\right)+\dot{Q}_{evap}-\dot{Q}_{conv}$$ (41)

where, $\dot{Q}_{evap}$ and $\dot{Q}_{conv}$ (kW) are the evaporation and convection heat transfer rates between the drying air and wood, which can be calculated as follows:

$$\dot{Q}_{evap}=r\dot{m}_{wv,s}A_{surf}$$ (42)

$$\dot{Q}_{conv}=hA\left(T_a-T_{SO}\right)$$ (43)

The specific water vapor mass flow rate ($\dot{m}_{wv,surf}$) (kg/m^2 s) to the drying air can be calculated as follows:

$$\dot{m}_{wv,surf} = \frac{h_D}{R_{wv}T_{SO}}\left(P_{wv,surf} - P_{wv,a}\right)$$ (44)

The vapor pressure on the wood surface can be determined from the sorption isotherms of wood. The mass transfer coefficient (h_D) (m/s) can be calculated from the convection heat transfer coefficient (h) (kW/m^2 K) as follows:

$$h_D = h\frac{1}{\rho_{a,mt}cp_{a,mt}Le^{0.58}}\bigg/\left(1 - \frac{\rho_{wv,m}}{P}\right)$$ (45)

9.6.6 Water Transfer Model above FSP

Water transfer in wood involves liquid free water and water vapor flow while MC of lumber is above the FSP.

According to Darcy's law the liquid free water flux is in proportion to pressure gradient and permeability. So Darcy's law for liquid free water may be written as:

$$J_f = \frac{K_l\rho_l}{\mu_l}\cdot\frac{\partial P_c}{\partial \chi}$$ (46)

where,

J_f = liquid free water flow flux, kg/m^2·s,

K_l = specific permeability of liquid water, $m^3(liquid)/m$,

ρ_l = density of liquid water, kg/m^3,

μ_l = viscosity of liquid water, p_a·s,

P_c = capillary pressure, p_a,

χ = water transfer distance, m,

$\partial p_c/\partial \chi$ = capillary pressure gradient, p_a/m.

The water vapor flow flux is also proportional to pressure gradient and permeability as follows:

$$J_{vf} = \frac{K_V\rho_v}{\mu_V}\cdot\frac{\partial P_V}{\partial \chi}$$ (47)

where,

J_{vf} = water vapor flow flux, kg/m^2·s,

K_V = specific permeability of water vapor, $m^3(vapor)/m$,

ρ_v,μ_v = density and viscosity of water vapor respectively, kg/m^3 and p_a·s,

$\partial p_V/\partial \chi$ = vapor partial pressure gradient, p_a/m.

Therefore, the water transfer equation above FSP during high temperature drying can be written as:

$$\rho_s \frac{\partial(MC)}{\partial t} = \frac{\partial}{\partial x}\left(J_f + J_{vf}\right) \tag{48}$$

where,

ρ_s = basic density of wood, kg/m^3,

MC = moisture content of wood, %,

t = time, s,

$\partial(MC)/\partial t$ = the rate of moisture content change, %/s,

x = water transfer distance, m.

9.6.7 Water Transfer Model below FSP

Water transfer in wood below FSP involves bound water diffusion and water vapor diffusion. The bound water diffusion in lumber usually is unsteady diffusion: the diffusion equation follows Fick's second law as follows:

$$\frac{\partial(MC)}{\partial t} = \frac{\partial}{\partial x}\left(D_b \frac{\partial(MC)}{\partial x}\right) \tag{49}$$

where D_b is bound water diffusion coefficient, m^2/s, $\partial(MC)/\partial x$ is MC gradient of lumber, %/m.

The bound water diffusion flux J_b can be expressed as:

$$J_b = D_b \rho_s \frac{\partial(MC)}{\partial x} \tag{50}$$

where,

ρ_s is basic density of wood, kg/m^3.

The water vapor diffusion equation is similar to bound water diffusion equation as follows:

$$\frac{\partial(MC)}{\partial t} = \frac{\partial}{\partial(MC)}\left(D_V \frac{\partial(MC)}{\partial x}\right) \tag{51}$$

where, D_V is water vapor diffusion coefficient, m^2/s.

The water vapor diffusion flux can be expressed as:

$$J_V = D_V \rho_s \frac{\partial(MC)}{\partial x} \tag{52}$$

Therefore, the water transfer equation below FSP during high temperature drying can be expressed as:

$$\rho_s \frac{\partial(MC)}{\partial t} = \frac{\partial}{\partial x}(J_b + J_V) \qquad (53)$$

9.7 EXPERIMENTAL

Two types of wood samples (namely: Guilan spruce and pine) were selected for drying investigation. Natural defects such as knots, checks, splits, and so on which would reduce strength of wood are avoided. All wood samples were dried to a moisture content of approximately 30%. The effect of drying temperature and drying modes on the surface roughness, hardness, and color development of wood samples are evaluated.

9.7.1 Surface Roughness

The average roughness is the area between the roughness profile and its mean line, or the integral of the absolute value of the roughness profile height over the evaluation length:

$$R_a = \frac{1}{L}\int_0^L |r(x)dx| \qquad (54)$$

When evaluated from digital data, the integral is normally approximated by a trapezoidal rule:

$$R_a = \frac{1}{N}\sum_{n=1}^N |r_n| \qquad (55)$$

The root-mean-square (RMS) average roughness of a surface is calculated from another integral of the roughness profile:

$$R_q = \sqrt{\frac{1}{L}\int_0^L r^2(x)dx} \qquad (56)$$

The digital equivalent normally used is:

$$R_q = \sqrt{\frac{1}{N}\sum_{n=1}^N r_n^2} \qquad (57)$$

R_z (ISO) is a parameter that averages the height of the five highest peaks plus the depth of the five deepest valleys over the evaluation length. These parameters which are characterized by ISO 4287 were employed to evaluate influence of drying methods on the surface roughness of the samples [21-23].

We investigated the influence of drying temperatures on the surface roughness characteristics of veneer samples as well. The results showed that the effect of drying temperatures used in practice is not remarkable on surface roughness of the sliced veneer and maximum drying temperature (130°C) applied to sliced veneers did not affect significantly surface roughness of the veneers. Veneer sheets were classified into four groups and dried at 20, 110, 150, and 180°C. According to the results, the smoothest surfaces were obtained for 20°C drying temperature while the highest values of surface roughness were obtained for 180°C. Because some surface checks may develop in the oven drying process. It was also found in a study that the surface roughness values of beech veneers dried at 110°C was higher than that of dried at 20°C.

In another experimental study, veneer sheets were oven dried in a veneer dryer at 110 (normal drying temperature) and 180°C (high drying temperature) after peeling process. The surfaces of some veneers were then exposed at indoor laboratory conditions to obtain inactive wood surfaces for glue bonds, and some veneers were treated with borax, boric acid and ammonium acetate solutions. After these treatments, surface roughness measurements were made on veneer surfaces. Alder veneers were found to be smoother than beech veneers. It was concluded that the values mean roughness profile (R_a) decreased slightly or no clear changes were obtained in R_a values after the natural inactivation process. However, little increases were obtained for surface roughness parameters, no clear changes were found especially for beech veneers.

The changes created by weathering on impregnated wood with several different wood preservatives were investigated. The study was performed on the accelerated weathering test cycle, using UV irradiation and water spray in order to simulate natural weathering. Wood samples were treated with ammonium copper quat (ACQ 1900 and ACQ 2200), chromated copper arsenate (CCA), Tanalith E 3491 and Wolmanit CX-8 in accelerated weathering experiment. The changes on the surface of the weathered samples were characterized by roughness measurements on the samples with 0, 200, 400, and 600 hr of total weathering. Generally, the surface values of alder wood treated with copper-containing preservatives decreased with over the irradiation time except for treated Wolmanit CX-82% when comparing unweathered values. Surface values of pine treated samples generally increased with increasing irradiation time except for ACQ-1,900 groups.

Because the stylus of detector was so sensitive first each sample was smoothened with emery paper then measurement test was performed before and after drying. The Mitutoyo Surface roughness tester SJ-201P instrument was employed for surface roughness measurements. Cut-off length was 2.5 mm, sampling length was 12.5 mm, and detector tip radius was 5 μm in the surface roughness measurements. Table 3 and Table 2 displays the changes in surface roughness parameters (R_a, R_z and R_q) of the Pine and Guilan spruce at varying drying methods. In both cases the surface roughness becomes higher during microwave and infrared heating while surface smoothness of both pine and Guilan spruce increased during convection and combined drying. However, the roughness of wood is a complex phenomenon because wood is an anisotropic and heterogeneous material. Several factors such as anatomical differences, growing characteristics, machining properties of wood, pre-treatments (e.g. steaming, drying, etc.) applied to wood before machining should be considered for the evaluation of the surface roughness of wood.

TABLE 2 Surface roughness (µm) for pine.

Drying methods	Drying conditions	R_a	R_z	R_q
Microwave	Before drying	4.52	24.68	5.39
	After drying	5.46	30.21	6.62
Infrared	Before drying	4.42	25.52	5.43
	After drying	4.87	26.55	5.69
Convection	Before drying	4.66	26.87	5.86
	After drying	4.08	24.64	5.12
Combined	Before drying	5.23	32.59	6.42
	After drying	3.41	21.7	4.27

TABLE 3 Surface roughness (µm) for Guilan spruce.

Drying methods	Drying conditions	R_a	R_z	R_q
Microwave	Before drying	6.44	34.18	7.85
	After drying	7.77	44.3	9.82
Infrared	Before drying	4.92	30.61	6.30
	After drying	6.42	38.93	8.17
Convection	Before drying	4.97	32.41	6.5
	After drying	4.78	32.27	6.34
Combined	Before drying	10.41	59.5	13.37
	After drying	9.11	54.31	11.5

9.7.2 Hardness

Hardness represents the resistance of wood to indentation and marring. In order to measure the hardness of wood samples, the Brinell hardness method was applied. In this method, a steel hemisphere of diameter 10 mm was forced into the surface under test. The Brinell method measures the diameter of the mark caused by the steel ball in the specimens. The specimens were loaded parallel and perpendicular to the direction of wood grains. After applying the force the steel ball was kept on the surface for about 30 s. The values of hardness are shown in Figures 2 and 3 respectively. In both type of samples the hardness measured in longitudinal direction is reported to be higher than tangential. The amount of fibers and its stiffness carrying the load are expected to be lower when the load direction is angled to the grain. Results showed that hardness of wood increased in combined drying. The hardness of wood is proportional to its density. The hardness of wood varies, depending on the position of the measurement. Latewood is harder than early wood and the lower part of a stem is harder than the upper part. Increase in moisture content decreases the hardness of wood. It was observed the effect of different drying temperatures during air circulation drying. The result indicates no significant influence of temperature on hardness: still the specimens dried at higher temperature gave a hard and brittle impression. It was also investigated whether wood hardness is affected by temperature level during microwave drying and whether the response is different from that of conventionally dried wood. It was concluded that there is a significant difference in wood hardness parallel to the grain between methods when drying progresses to relatively lower level of moisture content that is wood hardness becomes higher during microwave drying. Variables such as density and moisture content have a greater influence on wood hardness than does the drying method or the drying temperature [24].

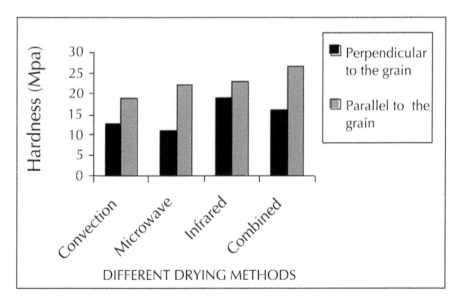

FIGURE 2   Brinell hardness for Guilan spruce.

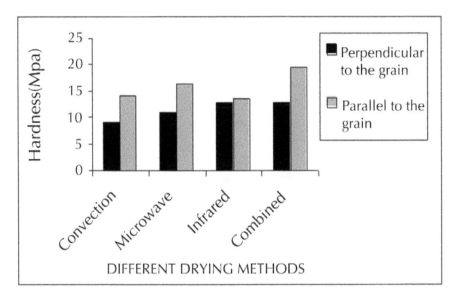

FIGURE 3 Brinell hardness for pine.

9.7.3 Color Development Measurement

Color development of wood surfaces can be measured by using optical devices such as spectrophotometers. With optical measurement methods, the uniformity of color can be objectively evaluated and presented as L*, a*, and b* coordinates named by CIEL*a*b* color space values. Measurements were made both on fresh and dried boards and always from the freshly planted surface. Three measurements in each sample board were made avoiding knots and other defects and averaged to one recording. The spectrum of reflected light in the visible region (400–750 nm) was measured and transformed to the CIEL*a*b* color scale using a 10° standard observer and D65 standard illuminant.

These color space values were used to calculate the total color change (ΔE^*) applied to samples according to the following equations:

$$
\begin{aligned}
\Delta L^* &= L_f^* - L_i^* \\
\Delta a^* &= a_f^* - a_i^* \\
\Delta b^* &= b_f^* - b_i^* \\
\Delta E^* &= \sqrt{(\Delta L^*)^2 + (\Delta a^*)^2 + (\Delta b^*)^2}
\end{aligned}
\tag{58}
$$

f and i are subscripts after and before drying respectively.

In this 3D coordinates, L* axis represents non-chromatic changes in lightness from an L* value of 0 (black) to an L* value of 100 (white), + a* represents red, –a* represents green, + b* represents yellow and –b* represents blue.

As can be seen from Figure 4 and Figure 5 color space values of both pine and Guilan spruce changed after drying.

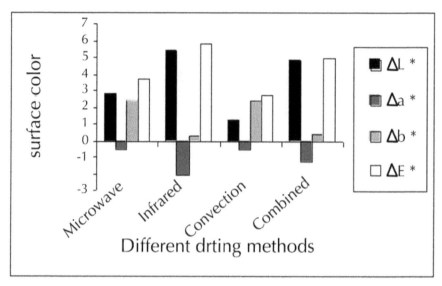

FIGURE 4   Surface color of pine.

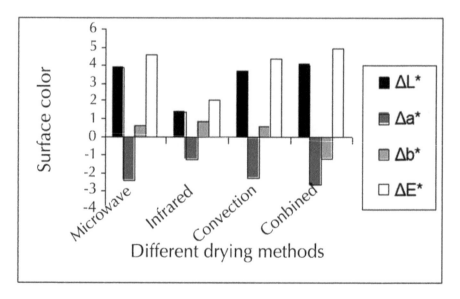

FIGURE 5   Surface color.

Results shows that Δa^* generally decreased but Δb^* increased for both pine and Guilan spruce wood samples except during combined heating. The lightness values ΔL^* increased during drying. The L^* of wood species such as tropical woods which originally have dark color increases by exposure to light. This is due to the special species and climate condition of pines wood. Positive values of Δb^* indicate an increment of yellow color and negative values an increase of blue color. Negative values of Δa^* indicate a tendency of wood surface to greenish. A low ΔE^* corresponds to a low color change or a stable color. The biggest changes in color appeared in ΔE^* values of pine samples during infrared drying while for Guilan spruce it was reversed. Due to differences in composition of wood components, the color of fresh, untreated wood varies between different species, between different trees of the same species and even within a tree. Within a species wood color can vary due to the genetic factors and environmental conditions. In discoloration, chemical reactions take place in wood, changing the number and type of chromophores.

Discolorations caused by the drying process are those which actually occur during drying and are mainly caused by non-microbial factors. Many environmental factors such as solar radiation, moisture, and temperature cause weathering or oxidative degradation of wooden products during their normal use; these ambient phenomena can eventually change the chemical, physical, optical, and mechanical properties of wood surfaces.

A number of studies have been conducted that have attempted to find a solution to kiln brown stain, the majority of them being pre-treatment processes. Biological treatment, compression rolling, sap displacement, and chemical inhibitors have been used as pre-treatments [25].

In all cases these processes were successful in reducing or eliminating stain but were not considered economically viable. Vacuum drying and modified schedules have been tried as modified drying processes with only limited success. Within industry various schedules have been developed, though these are generally kept secret and it is difficult to gauge their success. Generally, it seems that industry has adopted a post-drying process involving the mechanical removal of the kiln brown stain layer.

9.8 CONCLUSION

Microwave processing of materials is a relatively new technology that provides new approaches to improve the physical properties of materials. Microwave drying generate heat from within the grains by rapid movement of polar molecules causing molecular friction and help in faster and more uniform heating than does conventional heating. If wood is exposed to an electromagnetic field with such high frequency as is characteristic for microwaves, the water molecules, which are dipoles, begin to turn at the same frequency as the electromagnetic field. Wood is a complex composite material, which consists mainly of cellulose (40–45%), hemicelluloses (20–30%), and lignin (20–30%). These polymers are also polar molecules, and therefore even they are likely to be affected by the electromagnetic field. This could possibly cause degradation in terms wood hardness. For Guilan spruce the average of hardness is shown to be much higher than pine. From the experimental results it can be observed that in combined microwave dryer, the hardness was relatively improved in comparison to the other drying methods. Microwave and infrared drying can increase wood surface roughness while the smoothness of wood increases during convection and combined drying. The effect varies with the wood species. Thus, this work suggests keeping the core temperature below the critical value until the wood has dried below fiber saturation as one way of ensuring that the dried wood is acceptably bright and light in color.

KEYWORDS

- **Capillary pressure**
- **Drying rate**
- **Fiber saturation point**
- **Mass conservation equations**
- **Mass transfer coefficients**
- **Non-hygroscopic**

REFERENCES

1. Incropera, F. P. and Dewitt, D. P. *Fundamentals of Heat and Mass Transfer*. Second ed., Wiley, New York (1985).
2. Ghali, K., Jones, B., and Tracy, E. Modeling Heat and Mass Transfer in Fabrics. *Int. J. Heat Mass Transfer*, **38**(1), 13–21 (1995).
3. Flory, P. J. *Statistical Mechanics of Chain Molecules*. Interscience Pub., New York (1969).
4. Hadley, G. R. Numerical Modeling of the Drying of Porous Materials. In *Proceedings of the Fourth International Drying Symposium, Vol. 1.*, pp. 151–158 (1984).
5. Hong, K., Hollies, N. R. S., and Spivak, S. M. Dynamic Moisture Vapor Transfer through Textiles, Part I: Clothing Hygrometry and the Influence of Fiber Type. *Textile Res. J.* **58**(12), 697–706 (1988).
6. Chen, C. S. and Johnson, W. H. Kinetics of Moisture Movement in Hygroscopic Materials, In Theoretical Considerations of Drying Phenomenon. *Trans. ASAE.*, **12**, 109–113 (1969).
7. Barnes, J. and Holcombe, B. Moisture Sorption and Transport in Clothing during Wear. *Textile Res. J.*, **66**(12), 777–786 (1996).
8. Chen, P. and Pei, D. A Mathematical Model of Drying Process. *Int. J. Heat Mass Transfer*, **31**(12), 2517–2562 (1988).
9. Davis, A. and James, D. Slow Flow through a Model Fibrous Porous Medium. *Int. J. Multiphase Flow*, **22**, 969–989 (1996).
10. Jirsak, O., Gok, T., Ozipek, B., and Pau, N. Comparing Dynamic and Static Methods for Measuring Thermal Conductive Properties of Textiles. *Textile Res. J.*, **68**(1), 47–56 (1998).
11. Kaviany, M. *Principle of Heat Transfer in Porous Media*. Springer, New York (1991).
12. Jackson, J. and James, D. The Permeability of Fibrous Porous Media. *Can. J. Chem. Eng.*, **64**, 364–374 (1986).
13. Dietl, C. and George, O. P., and Bansal, N. K. Modeling of Diffusion in Capillary Porous Materials during the Drying Process. *Drying Technol*, **13** (1, 2), 267–293 (1995).
14. Ea, J. Y. *Water Vapor Transfer in Breathable Fabrics for Clothing*. PhD thesis, University of Leeds (1988).
15. Haghi, A. K. Moisture permeation of clothing. *JTAC*, 76, 1035–1055(2004).
16. Haghi, A. K. Thermal analysis of drying process. *JTAC*, **74**, 827–842 (2003).

17. Haghi, A. K. *Some Aspects of Microwave Drying.* The Annals of Stefan cel Mare University, Year VII, No. 14, pp. 22–25 (2000).
18. Haghi, A. K. A Thermal Imaging Technique for Measuring Transient Temperature Field. *An Experimental Approach.* The Annals of Stefan cel Mare University, Year VI, No. 12, pp. 73–76 (2000).
19. Haghi, A. K. Experimental Investigations on Drying of Porous Media using Infrared Radiation. *Acta Polytechnica,* **41**(1), 55–57 (2001).
20. Haghi, A. K. A Mathematical Model of the Drying Process. *Acta Polytechnica,* **41**(3), 20–23 (2001).
21. Haghi, A. K. Simultaneous Moisture and Heat Transfer in Porous System. *Journal of Computational and Applied Mechanics,* **2**(2), 195–204 (2001).
22. Haghi, A. K. A Detailed Study on Moisture Sorption of Hygroscopic Fiber. *Journal of Theoretical and Applied Mechanics,* **32**(2), 47–62 (2002).
23. Dietl, C. and George, O. P., and Bansal, N. K. Modeling of Diffusion in Capillary Porous Materials during the Drying Process. *Drying Technol,* **13**(1, 2), 267–293 (1995).
24. Ea, J. Y. *Water Vapor Transfer in Breathable Fabrics for Clothing.* PhD thesis, University of Leeds (1988).
25. Flory, P. J. *Statistical Mechanics of Chain Molecules.* Interscience Pub., New York (1969).

Index

P

Milton Keynes UK
Ingram Content Group UK Ltd.
UKHW031145141024
449569UK00024B/1046

9 781774 632628